SpringerBriefs in Applied Sciences and Technology

SpringerBriefs present concise summaries of cutting-edge research and practical applications across a wide spectrum of fields. Featuring compact volumes of 50–125 pages, the series covers a range of content from professional to academic.

Typical publications can be:

- A timely report of state-of-the art methods
- An introduction to or a manual for the application of mathematical or computer techniques
- A bridge between new research results, as published in journal articles
- A snapshot of a hot or emerging topic
- An in-depth case study
- A presentation of core concepts that students must understand in order to make independent contributions

SpringerBriefs are characterized by fast, global electronic dissemination, standard publishing contracts, standardized manuscript preparation and formatting guidelines, and expedited production schedules.

On the one hand, **SpringerBriefs in Applied Sciences and Technology** are devoted to the publication of fundamentals and applications within the different classical engineering disciplines as well as in interdisciplinary fields that recently emerged between these areas. On the other hand, as the boundary separating fundamental research and applied technology is more and more dissolving, this series is particularly open to trans-disciplinary topics between fundamental science and engineering.

Indexed by EI-Compendex and Springerlink.

More information about this series at http://www.springer.com/series/8884

Hanhua Zhu · Zhijun Wu
Mengchong Chen · Yongli Zhao

Controlling Differential Settlement of Highway Soft Soil Subgrade

A New Method and Its Engineering Applications

ZHEJIANG UNIVERSITY PRESS
浙江大学出版社

 Springer

Hanhua Zhu
Highway Bureau of Zhejiang Provinice
Hangzhou, Zhejiang
China

Zhijun Wu
School of Civil Engineering
Wuhan University
Wuhan, Hubei
China

Mengchong Chen
Ningbo Communications Planning
　Institute Co., Ltd.
Ningbo
China

Yongli Zhao
Ningbo Communications Planning
　Institute Co., Ltd.
Ningbo
China

ISSN 2191-530X　　　　　　ISSN 2191-5318　(electronic)
SpringerBriefs in Applied Sciences and Technology
ISBN 978-981-13-0721-8　　　ISBN 978-981-13-0722-5　(eBook)
https://doi.org/10.1007/978-981-13-0722-5

Jointly published with the Zhejiang University Press, Hangzhou, China

The print edition is not for sale in China Mainland. Customers from China Mainland please order the print book from: Zhejiang University Press.

Library of Congress Control Number: 2018942162

Printed on acid-free paper

This Springer imprint is published by the registered company Springer Nature Singapore Pte Ltd. part of Springer Nature
The registered company address is: 152 Beach Road, #21-01/04 Gateway East, Singapore 189721, Singapore

Preface

Differential settlement in highway soft soil subgrade is commonly observed, which induces a series of problems such as bump at bridgehead. Up to now, many treatments and analysis methods of highway soft soil subgrade have been developed. However, there are still few successful cases, and the application of the aforementioned methods is limited. Why does it happen? It is generally accepted that the calculated stress of soft soil foundation is accurate whereas there is significant error in the deformation calculation, which contradicts with the function $y = f(x)$. The key matter is the correspondence between the mechanical analysis theory of highway soft soil subgrade and the actual application method. There are two points that are easily neglected: (1) The main difference between the continuum and discrete object of subgrade is that the discrete object can bear compressive force, but fails in bearing tensile force and moment. Instead, the continuum can not only bear pressure but also tension and moment. Therefore, it is important to the soft soil subgrade in a stable deformation state to maintain its continuity. (2) The soft soil foundation of soft soil subgrade has the characteristics of high pore ratio, high compressibility, high water content, low permeability, low strength, strong rheological property, strong structural property, high sensitivity, etc., which easily leads to the differential settlement of subgrade. Therefore, it is critical to control the unstable continuous settlement caused by the rheology of soft soil.

SEM results regarding the change of the soft soil microstructure during loading show that the soil particle structure varies with the variation of the applied stress while the consolidation and rheological properties of the soft soil are dependent on its applied stress level. Therefore, the results obtained from the analysis based on structure mechanics usually have poor applicability. The differential settlement problem of highway soft soil subgrade, especially bump at bridgehead, can only be solved using the control measures of structural deformation compatibility instead of the theory or assumption of structural deformation compatibility.

Issue related to differential settlement of highway soft soil subgrade belongs to plane strain problem—its mechanical characteristics conform to Terzaghi theory assumption, which is required to control the force and deformation state of highway soft soil subgrade in a stable state. Therefore, the synthetic method should be used to solve the differential settlement problem of highway soft soil subgrade. In fact, the existing mechanical analysis and treatment methods of highway soft soil subgrade all imply the deformation compatibility assumption, which enables the stress and deformation state of the subgrade to be controlled. Moreover, the pressure, the tension, and the moment can be withstood. In practice, the stress and deformation stability of the subgrade is often neglected. For example, the slag subgrade can only bear pressure but fail to control the differential settlement of soft soil subgrade. The influence of consolidation and secondary consolidation of soft soil foundation is generally considered, whereas the unstable continuous settlement caused by the rheology of soft soil is often neglected. Consequently, it is not surprising that the differential settlement of soft soil subgrade especially the bump at bridgehead phenomenon is prominent.

This book aims to solve the two problems existing in the analysis theory and treatment methods of highway soft soil subgrade, and to put emphasis on the problem of matching the theory of mechanical analysis and the practical application method in highway soft soil subgrade. Based on the practical experiences and dialectical thinking, this monograph focuses on the indicators—including the stress and deformation state of highway soft soil subgrade, as well as unstable continuous settlement caused by soft soil rheology—by using the structural deformation compatibility control method to improve the design method of soft soil subgrade. Based on the design standard, more attention is given to the control of the deformation stability of highway soft soil subgrade and the longitudinal transition. The key technology is the control of the plate or frame used in the highway soft soil subgrade to reduce differential settlement and the longitudinal transitional pile.

Among them, the length of the longitudinal transitional pile and the gradually varying height of the bridgehead subgrade are highly relevant. The lightweight materials need to be controlled according to the hierarchical stability and vehicle load distribution structure. Controlling the differential settlement of bridge foundation in the allowable range can solve the problem of differential settlement of highway soft soil subgrade (including bump at bridgehead).

I would like to give thanks to the many friends who made important contributions to the preparation of this book. In particular, I would like to pay tribute to academician Sun Jun for the careful guidance provided. Finally, we sincerely thank the Fundamental Research Funds for the Central Universities for the financial support.

Because of the limited level of the author, errors inevitably exist in this book. As such, any constructive criticism is welcome.

Hangzhou, China   Hanhua Zhu
Wuhan, China   Zhijun Wu
Ningbo, China   Mengchong Chen
Ningbo, China   Yongli Zhao
January 2018

Contents

Summary

In general, when the force and displacement state of the structure can be divided without affecting its stability, the analytical method should be used. When the force and displacement state of the structure cannot be divided otherwise its stability is affected, the synthetic method should be used. Issue related to differential settlement of highway soft soil subgrade belongs to the plane strain problem—its mechanical characteristics conform to Terzaghi theory assumption. It is very important to control the force and deformation state of the highway soft soil subgrade. Usually, the synthetic method should be adopted to solve the differential settlement problems in highway soft soil subgrade. In terms of stable highway soft soil subgrade, it is critical to control the unstable continuous settlement caused by soft soil rheology. The initial rheological stress value of soft soil is similar to that of the unit buoyancy coefficient of water. When the subgrade bottom pressure is lower than the initial value of soft soil rheological stress, it is enough to control the force and deformation state stability of soft soil subgrade or using deck plate. Otherwise, an additional adoption of the pile foundation for transition is essential. According to the structure deformation coordination control method, the soft soil subgrade design method is improved in this paper. The aforementioned two key parameters, i.e., the force and deformation stability of soft soil subgrade and initial rheological stress value for soft soil should be well controlled on the basis of design specifications. The developed technology guarantees a stable state of the highway soft soil subgrade. In addition, a new design method for dealing with the differential settlement problems of highway soft soil (including bump at bridgehead) is proposed. With this new method, the influence caused by the differential settlement of highway soft soil subgrade (including bump at bridgehead) is well solved. This book can be used as a reference for the scientific researchers and technical staff engaging in the field of highway engineering, and can also be used as a reference for teachers and students of related majors in colleges.

Part I
Improved Design Method of Highway Soft Soil Subgrade

Targeting at the two aforementioned problems existing in analysis theory and treatment methods of highway soft soil subgrade, the compatibility between the highway soft soil subgrade mechanical analysis theory and its practical application is grasped in this monograph. It is very important to control the stability of stress and deformation state of the subgrade, as well as to refer to soft soil initial rheology stress value. Therefore, in accordance with the structural deformation compatibility control method, the design method of soft soil subgrade is improved. In addition, on the basis of design specification, two key parameters, i.e., the stability of subgrade stress and deformation state, and the initial rheology stress value of soft soil are demonstrated with a better understanding. A series of technologies are developed to guarantee a stable balanced state of the highway soft soil subgrade is achieved, and new design method on effectively treating differential settlement of highway soft soil subgrade (including bumping at bridgehead) is also developed. Differential settlement of highway soft soil is controlled in an allowable range, and its influences including bumping at bridgehead are reduced [1–7].

References

1. K. Terzaghi, *Theoretical Soil Mechanics* (Wiley, New York, 1943)
2. J. Sun, *The Rheology and Engineering Applications of Geotechnical Materials* (China Architecture & Building Press, Beijing, 1999). (In Chinese)
3. X. Gong, Thinking of numerical analysis of geotechnical engineering. Rock Soil Mech. **32**(2), 321–325 (2011). (In Chinese)
4. J. Sun, *Collected Papers of Academician Sun Jun at the 80th Birthday* (Tongji University Press, Shanghai, 2006). (In Chinese)
5. B. Liu, *Collected Papers of Liu Baochen* (Central South University Press, Changsha, 2011). (In Chinese)
6. H. Zhu et al., *Engineering Structural Stability, Balance and Deformation Compatibility Control Measures and Applications* (China Communications Press Co., Ltd., Beijing, 2015). (In Chinese)
7. H. Zhu, Z. Zhou, *Thinking of the Issues Concerning Civil Engineering Structure Force Safety* (China Communications Press, Beijing, 2012). (In Chinese)

Chapter 1
Analysis of Differential Settlement of Highway Soft Soil Subgrade

The soft soil subgrade has the characteristics of large pore ratio, high compressibility, high water content, low permeability, low strength, strong rheological property, strong structural property, and high sensitivity, allowing it to easily lead to differential settlement of subgrade. The core problem is the inconsistency of the designed analysis model (Fig. 1.1a) of highway soft soil subgrade and actual stress and deformation state of soft soil slag road subgrade (Fig. 1.1b, c).

The ideal subgrade belongs to the continuum model, whereas the slag road subgrade belongs to dispersion. The main difference between the continuum and the dispersion is that the discrete objects could withstand pressure, but is unable to bear tension and moment, while a continuum can withstand pressure, tension, and moment.

In practical engineering, the vehicle load is a typical dynamic load (according to the statistics in the port of some area overloaded container vehicles weighing 200 t), where the tensile and bending zone form in the bottom of the subgrade. For Fig. 1.1b, it does not display a pure pressure zone under ideal conditions (as shown in Fig. 1.1a). At this point, if the vehicle is treated as a uniformly distributed load according to current specifications, a continuum model (Fig. 1.1a) will be used, leading to significant great errors.

At present, many measures have been proposed to achieve the stability of the stress and deformation of highway subgrade, including the adoption of EPS, foam concrete, and spacer plates. Among the aforementioned measures, EPS and foam concrete not only change the stability of the stress and deformation state, but also change the load size, whereas the spacer plate only changes the stability of the stress and deformation state of the subgrade. Taking a narrow bridge project of 104 National Road as an example, the effect of the stability of subgrade stress and deformation state are qualitatively illustrated, based on which the quantitative research is later given.

The narrow bridge project of State Road 104 west line is an old bridge broadening project—with the height of abutment landfill of 2.4 m and the soft soil layer thickness

© Springer Nature Singapore Pte Ltd. and Zhejiang University Press 2019
H. Zhu et al., *Controlling Differential Settlement of Highway Soft Soil Subgrade*,
SpringerBriefs in Applied Sciences and Technology,
https://doi.org/10.1007/978-981-13-0722-5_1

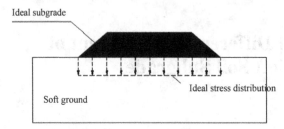

(a) Ideal stress distribution of highway soft soil subgrade (designed analysis model)

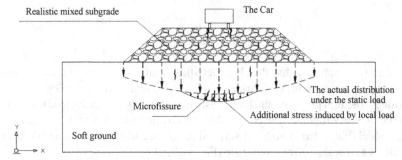

(b) Actual stress and deformation state under subgrade static load and vehicle local dynamic
load in slag road subgrade of highway soft soil (different to designed analysis model)

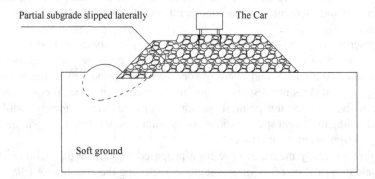

(c) Local lateral slippage diagram caused by slag road subgrade of highway soft soil
(different to designed analysis model)

Fig. 1.1 Contrast on designed model of highway soft soil subgrade and actual state of slag road subgrade

of 39.5 m. The old bridge was opened to traffic in 1999, and the broadened new bridge was opened to traffic in 2009. The traditional lap board technology was adopted when building the old bridge, while slag was directly filled on the spacer plate when in broadening the bridge. The bridgehead subgrade of the old bridge is still in settlement,

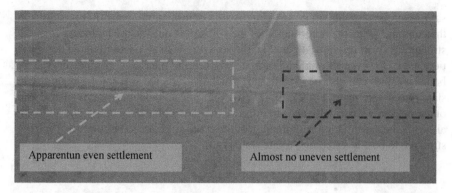

Fig. 1.2 Contrast of the old road and broadened road surfaces of State Road west line broaden project

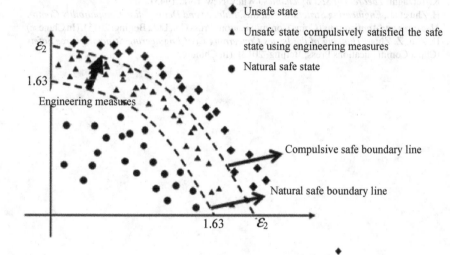

Fig. 1.3 Safety curve of soft soil stratum engineering

and the settlement value has exceeded that of the broadened bridgehead soft subgrade, with a total settlement of 11.23 cm. After 4 years of operation, the total settlement of the bridgehead subgrade in the broadened new bridge was 4.75 cm with no obvious differential settlement observed (as shown in Fig. 1.2).

The above example reveals that only a deep analysis of the differential settlement of highway soft soil subgrade can then reach an effective treatment. Through a large number of mechanical tests on the soft soil sampled obtained from the highway, the safety curve of the soft soil layer of No. 2 road is obtained, as illustrated in Fig. 1.3.

According to the rheological properties of the silt soft soil sample, the safe boundary line under the normal stress of 50 MPa is drawn (as shown in Fig. 1.3). When the deformation of the soft soil is within the boundary line, the structure is in a natural safe state, namely the whole structure is in a deformation compatibility state. On the contrary, when the deformation of the soft soil is outside the boundary line, the structure is in an unsafe state, that is, the structure is in an uncoordinated deformation state. However, through appropriate engineering measures, compulsory safety boundary lines can also be established, which converts the soft soil from an unsafe state into a safe state. For example, grouting, pile foundation, spacer plates, and other measures are usually used to improve the safety boundary line of the soft soil, allowing the deformation coordination of the entire structure to be controlled and the structure to be converted into a safe state [1–3].

References

1. K. Terzaghi, *Theoretical Soil Mechanics* (Wiley, New York, 1943)
2. H. Zhu et al., *Engineering Structural Stability, Balance and Deformation Compatibility Control Measures and Applications* (China Communications Press Co., Ltd., Beijing, 2015). (In Chinese)
3. H. Zhu, Z. Zhou, *Thinking of the Issues Concerning Civil Engineering Structure Force Safety* (China Communications Press, Beijing, 2012). (In Chinese)

Chapter 2
Overview of Methods for Treating the Differential Settlement of the Soft Soil Highway Subgrade (Including the Bumps at Bridgeheads)

With the rapid construction of highways in China since the 1990s, methods for treating the differential settlement of the soft soil highway subgrade (including the bumps at bridgeheads) have gradually been developed. In particular, successful application of a lime-treated soil subgrade has promoted experimentation and practice of basic theories concerning the treatment of differential settlement of the soft soil highway subgrade (including the bumps at bridgeheads). In this regard, the Department of Transportation of one province specially developed the *Implementation Plan for Treatment of Three Common Major Quality Problems in Highway Projects* in 1999. Furthermore, many technical solutions for treating the differential settlement of the soft soil highway subgrade (including the bumps at bridgeheads) have been developed in the transportation system throughout the province to improve the treatment of bumps at bridgeheads. For example, the first layer of the soft soil subgrade at the bridgehead of the transit highway in western Wenzhou City was paved with large stones in 1999. This treatment significantly improved the differential settlement of the subgrade at the bridgeheads. With respect to a 20 m three-span bridge from Jiaxing to Suzhou, where the bridgehead subgrades at the Jiaxing end and Suzhou end were paved with a soil–stone slag mixture and lime-treated soil, respectively, a bump at the bridgehead occurred at the Jiaxing end but was not detected at the Suzhou end in 2007, leading people to further realize the importance of stabilizing the state of the forced deformation of the soft soil highway subgrade.

When an old bridge on National Highway 104 (G104) in Huangyan was reconstructed and widened in 2009, Lane 3–Lane 4 were dealt with using lower diaphragm plate technology. As a result, there was no bump at the bridgeheads (Figs. 1.2 and 2.1). However, Lane 1–Lane 2, which were completed in 1999, were paved with a soil–stone slag mixture. As a result, a bump at the bridgehead was observed (as shown in Figs. 1.2 and 2.1).

© Springer Nature Singapore Pte Ltd. and Zhejiang University Press 2019
H. Zhu et al., *Controlling Differential Settlement of Highway Soft Soil Subgrade*,
SpringerBriefs in Applied Sciences and Technology,
https://doi.org/10.1007/978-981-13-0722-5_2

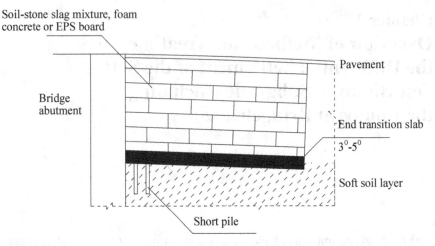

Soil-stone slag mixture, foam
concrete or EPS board

Bridge
abutment

Pavement

End transition slab

3^0-5^0

Soft soil layer

Short pile

Fig. 2.1 The structure of the subgrade in the project to widen the Second Ring West Lane of the National Highway

In 2012, the soft soil subgrade section of a coastal high-grade highway in one province was designed in compliance with the standards except that the subgrade was paved with a soil–stone slag mixture. However, despite what standardized construction used, the stability of the state of the forced deformation of the subgrade was not controlled and cracks developed in the base stratum of the pavement. These problems were solved by grouting reinforcement of the subgrade. This example facilitates the complete understanding of the importance of stabilizing the state of the forced deformation of the soft soil highway subgrade (Fig. 2.2).

In 2014, the project team continued to carry out research on the methods for treating the differential settlement of the soft soil highway subgrade (including the bumps at bridgeheads). Under the guidance of the academician Sun Jun as well as based on the research achievements of predecessors, some research was carried out mainly from the perspectives of the stability of the state of forced deformation in the soft soil subgrade and the consolidation and rheological properties of the soft soil subgrade. The following achievements were made:

(1) Understanding at the level of mechanics: adoption of mechanics theory for investigating the differential settlement of the soft soil highway subgrade (including the bumps at bridgeheads) is subject to a number of conditions. For example, the design that involves the stability of the state of forced deformation in the structure of the soft soil subgrade is very important; otherwise, some error is unavoidable.

(2) Understanding at the practical level: consider an analogy: The stability in making bean curd differs from the stability in pressing bean dregs; their results also differ.

Fig. 2.2 Standardized construction of the soft soil subgrade of a coastal high-grade highway in one province, and the cracks in its base stratum

(a) Lime-treated soil subgrade

(b) Light filler subgrade

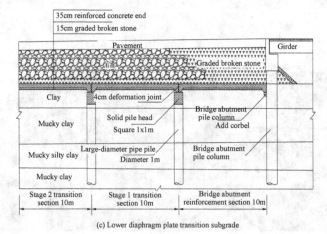

(c) Lower diaphragm plate transition subgrade

Fig. 2.3 Several methods for effectively treating the differential settlement of the soft soil highway subgrade (including the bumps at bridgeheads)

(3) The effect of pilot work in projects: the project team used lime-stabilized soil in the subgrade of seven highway projects in Jiaxing. As a result, good effects were achieved and no bumps formed at the bridgeheads. In these projects, 44.452 km of construction were completed, 2,091,000 m^3 of lime-stabilized soil was employed, the rate of cost savings was 23.6%, and a sustainable amount of waste soil was recycled. Thus, the ecological environment was protected. The pilot work in Ningbo, Wenzhou, and Taizhou also produced excellent results: structural measures, such as a lower diaphragm plate, were implemented to effectively treat the differential settlement of the soft soil highway subgrade (including the bumps at bridgeheads) (Fig. 2.3) [1–14].

References

1. J. Sun, *The Rheology and Engineering Applications of Geotechnical Materials* (China Architecture & Building Press, Beijing, 1999). (In Chinese)
2. X. Gong, Thinking of numerical analysis of geotechnical engineering. Rock Soil Mech. **32**(2), 321–325 (2011). (In Chinese)
3. H. Zhu et al., *Engineering Structural Stability, Balance and Deformation Compatibility Control Measures and Applications* (China Communications Press Co., Ltd., Beijing, 2015). (In Chinese)
4. Industrial Standard of the People's Republic of China, *JTG D30-2004 Code for Design of Highway Subgrades* (China Communications Press, Beijing, 2004). (In Chinese)
5. D. Gao, *Review and Prospect of Geotechnical Engineering* (China Communications Press, Beijing, 2002). (In Chinese)
6. Y. Guo, et al., *New Techniques for Controlling the Post-construction Settlement Deformation in the Subgrades and Treatment of the Bumps at Bridgeheads*. Collected Papers of the 9th Academic Conference on Soil Mechanics and Geotechnical Engineering© (Tsinghua University Press, Beijing, 2003). (In Chinese)
7. G.B. Liu, S.M. Liao, *Geotechnical Engineering in Soft Ground* (Tongji University Press, Shanghai, 2001)
8. S.R. Lo, M.R. Karim, C.T. Gnanendran, Consolidation and creep settlement of embankment on soft clay: prediction versus observation, *Geotechnical Predictions and Practice in Dealing with Geohazards* (Springer, Dordrecht, Netherlands, 2013)
9. H.E.M. Hunt, Settlement of railway track near bridge abutments. ICE Proc.-Transp. **123**(1), 68–73 (1997)
10. C.L. Snow, C.R. Nickerson, Case study of EPS geofoam lightweight fill for settlement control at bridge approach embankment. Geotech. Eng. Transp. Projects **1**, 580–589 (2004)
11. Y. Demura, M. Matsuo, *Optimization of Foundation of Bridge on Soft Ground*, in Reliability and Optimization of Structural Systems: Proceedings of the sixth IFIP WG7.5 working conference on reliability and optimization of structural systems (Springer, Boston, 1994, 1995), pp. 112–119
12. Z.D. Cui, S.X. Ren, Prediction of long-term settlements of subway tunnel in the soft soil area. Nat. Hazards **74**(2), 1007–1020 (2014)
13. J.Y. Wu, The settlement behaviors of granular backfill materials for high speed rail embankment. Geotech. Eng. Transp. Projects **2**(126), 1584–1591 (2004)
14. J.L. Briaud, R.W. James, S.B. Hoffman, *Settlement of Bridge Approaches (The Bump at the End of the Bridge)* (Transportation Research Board, Washington, 1997)

Chapter 3
Inspirations from the Stability of Ancient Structures

3.1 Inspirations from the Structure of Zhenwu Pavilion in Guangxi

Zhenwu Pavilion at Rongxian, Guangxi (Fig. 3.1) has two distinctive features [1, 2]:

(1) Jinglue Platform was built in the Tang Dynasty (approximately 1000 years ago) and is located on river sand that was compacted by filling soil within a brick wall; neither hard stones nor solid reinforced cement were employed. Zhenwu Pavilion was built in the Tang Dynasty and is located on Jinglue Platform (sand pile). The pavilion has survived five earthquakes over more than 400 years since its foundation is protected by excellent seismic isolation. This method has been applied in buildings in recent years—first, a layer of thick sand is first paved beneath the foundation of some reinforced cement buildings, and second, a reinforced concrete earth net pedestal is poured.

(2) The Zhenwu Pavilion has 20 large stand columns, of which eight columns extend to support the entire load of this three-storey pavilion. The columns are connected by girders. Four brackets are used at the top of the columns and bear four wooden structures; these brackets powerfully support this pavilion. This structure is similar to the rational load-carrying structure of a bench. Four inner columns on the second floor sustain the heavy load of the upper floor, beam frame, supporting columns, pavilion tiles and cresting. However, the column foot is located 3 cm away from the ground, which conforms to the principle of a lever structure. As a result, the pavilion remains as strong as a rock after three typhoons.

Fig. 3.1 Zhenwu Pavilion in Guangxi

3.2 Inspirations from the Destruction of India's Jaisalmer Sandcastle by Wastewater

India's Jaisalmer Sandcastle is a desert castle with a history of more than 800 years. With the development of the tourism industry, many inns, hotels and entertainment venues have been built in the castle, and residents can enjoy convenient tap water. Due to the lack of well-enhanced drainage facilities, a significant amount of waste water flows through simple drainage ditches and constantly scours the sandcastle's foundation. Consequently, some sand and stone walls have collapsed. Originally, 99 intact fortresses were available on the walls; at present, numerous fortresses have been destroyed. The sandcastle was originally in a temporary and conditioned balanced state. However, this state has been destroyed by the substantial amount of waste water that has scoured the sandcastle's foundation, causing damage to its original balanced system (Fig. 3.2).

The comparison of the structural stability between India's Jaisalmer Sandcastle and the Zhenwu Pavilion in Guangxi, China, reveals a lack of stable walls around the desert foundation for fixing sand, the brick and stone structure is not as stable as the wood structure. A significant amount of waste water constantly scours the sandcastle's foundation, which causes damage to the sandcastle [1, 2].

Fig. 3.2 India's Jaisalmer Sandcastle

3.3 Inspirations from Hegui Earth Building in Fujian

The Hegui Earth Building in Fujian (Fig. 3.3) was built in the Tenth Year of Emperor Yong Zheng in the Qing Dynasty (1732). This building covers an area of 1547 m² and has five stories that encompass 21.5 m. With more than 200 pine piles and beddings, the Hegui Earth Building was constructed on marshland and remains strong and intact after more than 200 hundred years. It is reported that the site was not found to be marshland at the time of site selection; when one storey of the building was completed, the entire building slowly declined to marshy land like a sunken ship. To prevent the declination, the builders had to place at least 100 m³ of piles in a row on the sunken walls. After such treatment, the foundation was considered to be firm and then they rammed walls to build a five-storey building. After more than 200 years, the Hegui Earth Building remains as strong as iron and exceptionally firm [1, 2].

From the perspective of structural mechanics, its unique design has enabled the Hegui Earth Building to stand after a very long time; its design is analyzed as follows:

First, the stability of the state of structural forced deformation is controlled to guarantee the reasonable force distribution and total deformation compatibility in this earth building.

Second, pine piles are used to reinforce the soft soil foundation to improve the overall effective bearing capacity of the soft soil foundation.

Based on the principle of "unsinkable" for the Hegui Earth Building, the differential settlement of the existing soft soil highway subgrade is analyzed at the level of mechanics to develop basic principles for treating the soft soil highway subgrade:

Fig. 3.3 Hegui Earth Building in Fujian

(1) control the stability of the state of the soft soil subgrade's forced deformation;
(2) reduce the quality of the soft soil subgrade and improve the bearing capacity of the soft soil subgrade.

According to the theory of soil consolidation, which is commonly employed in the treatment of the soft soil subgrade, subject to load, excess pore water pressure will occur in the saturated soft soil subgrade. As time passes, the water in the soil pore is discharged, the excess pore water pressure gradually fade, and the effective stress in the soil gradually increases until the excess pore water pressure completely disappears and the deformation stabilizes. This process is called soil consolidation.

In 1925, Terzaghi described soil consolidation at the theoretical and mathematical levels in a relatively strict manner for the first time and developed the classical theory of soil consolidation.

Terzaghi's one-dimensional consolidation theory is based on the following assumptions:

(1) Soil is completely saturated and homogeneous;
(2) The compression of soil skeleton and seepage flow of pore water occurs only in one direction (generally in vertical direction);
(3) Soil particles and pore water cannot be compressed;
(4) The seepage flow of water in soil conforms to the Darcy law; soil permeability is unchanged during consolidation;
(5) Soil is a linear elastic body; its compressibility is unchanged during consolidation;

(6) Loads are instantaneously applied at a time;

(7) The boundary of a soil layer is completely permeable or completely imperme-
able.

According to Terzaghi's one-dimensional consolidation theory, after the excess
pore water pressure in soil completely disappears, the process of soil consolidation
ends and soil deformation will cease. However, as indicated by the results of many
indoor tests and field engineering tests, when the excess pore water pressure in soft
soil completely disappears and the effective stress is unchanged, soil deformation
does not stop and still continues to slowly develop with time. This deformation is
attributed to the rheological properties of soft soil. The rheological properties mean
that the deformation of soil increases with time even the effective stress remains
unchanged, which is the secondary time effect of soil deformation. Obviously, in
order to make the assumptions of the consolidation theory closer to the actual situation, the structural deformation compatibility control measure should be employed
to ensure that the stability of the state of the soft soil subgrade's forced deformation
conforms to Terzaghi's one-dimensional consolidation theory [1–8].

References

1. H. Zhu et al., *Engineering Structural Stability, Balance and Deformation Compatibility Control Measures and Applications* (China Communications Press Co., Ltd., Beijing, 2015). (In Chinese)
2. H. Zhu, Z. Zhou, *Thinking of the Issues Concerning Civil Engineering Structure Force Safety* (China Communications Press, Beijing, 2012). (In Chinese)
3. K. Terzaghi, *Theoretical Soil Mechanics* (Wiley, New York, 1943)
4. J. Sun, *The Rheology and Engineering Applications of Geotechnical Materials* (China Archi-tecture & Building Press, Beijing, 1999). (In Chinese)
5. Y. Demura, M. Matsuo, *Optimization of Foundation of Bridge on Soft Ground*, in Reliability and Optimization of Structural Systems: Proceedings of the sixth IFIP WG7.5 working conference on reliability and optimization of structural systems (Springer, Boston, 1994, 1995), pp. 112–119
6. Z.D. Cui, S.X. Ren, Prediction of long-term settlements of subway tunnel in the soft soil area. Nat. Hazards **74**(2), 1007–1020 (2014)
7. J.Y. Wu, The settlement behaviors of granular backfill materials for high speed rail embankment. Geotech. Eng. Transp. Projects **2**(126), 1584–1591 (2004)
8. J.L. Briaud, R.W. James, S.B. Hoffman, *Settlement of bridge approaches (The bump at the end of the bridge)* (Transportation Research Board, Washington, 1997)

Chapter 4
Explorations of the Reasonable Structure of the Soft Soil Highway Subgrade and Solutions

The traditional "apple falling site prediction" (the apple issue) and "leaf falling site prediction" (the leaf issue) are governed by the same prediction theory. However, the actual prediction methods differ; the simplest approach is to adopt the method used by man (childhood, adulthood, and old age) for maintaining motion stability to resolve the "leaf issue". Therefore, if the "leaf issue" is resolved in the "apple way", the difference between the calculation result and actual result for an engineering structure is large, and a safety hazard may occur. An exact solution for the "apple issue" exists, whereas the "leaf issue" involves a probability event. For the probability boundary conditions, only when control measure is adopted the accurate method can then be employed to resolve the issue; the key is the structural deformation compatibility control measure (Fig. 4.1).

Example 4.1 Two key processes exist for making bean curd (1) add gypsum (halide salt) into raw stock for curdling; (2) apply force with boards above and below the stock to create a certain shape. Even if bean dregs are compacted, they easily swell and may become loose in the presence of water; the key point is that no curdling material is added to the bean dregs. The key technique for making bean curb is simple. The core solution for inhomogeneous deformation of the soft soil highway subgrade consists of controlling irregular movements of soft soil particles. With reference to the method for making bean curb, the stability of the state of the subgrade's forced deformation is simply controlled.

Example 4.2 The traditional soft soil subgrade analysis theory adopts deterministic methods such as the use of a spring, dashpot and sliding block, to simulate the uncertain movement characteristics of the soft soil subgrade (particles). Internationally, it is widely believed that the force calculation concerning the soft soil subgrade is relatively accurate but with a large error in the deformation calculation, which conflicts with the functional relationship $y = f(x)$. Theoretical logic is rigorous but technical measures are not duly implemented; when regular movements of the soft soil particles are technically controlled, the stability of the state of the subgrade's forced

© Springer Nature Singapore Pte Ltd. and Zhejiang University Press 2019
H. Zhu et al., *Controlling Differential Settlement of Highway Soft Soil Subgrade*,
SpringerBriefs in Applied Sciences and Technology,
https://doi.org/10.1007/978-981-13-0722-5_4

(a) Apple issue

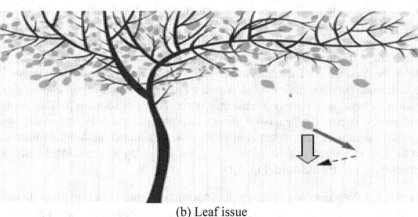

(b) Leaf issue

Fig. 4.1 Comparison of apple and leaf falling site prediction methods

deformation is simply controlled-the result of the soft soil subgrade calculation will fall within the permissible engineering range.

Example 4.3 Table 4.1 shows the results of a settlement comparison between a stable subgrade and an unstable subgrade. As indicated, the settlement of No. 1 highway's subgrade cannot be controlled, whereas the settlement of No. 2 highway's subgrade can be controlled. Their difference is attributed to the control of the stability of the state of the subgrade's forced deformation or structural deformation compatibility (Fig. 4.2).

According to Fig. 4.3, whereas the extent to which soft soil is subject to force differs, the soil particle structure differs; The consolidation and rheological properties of the soft soil subgrade correspond to the extent to which it is subject to force. Therefore, the soft soil highway subgrade involves the issue of plane strain, its mechanics

Table 4.1 Settlement comparison between a stable subgrade and an unstable subgrade

Indicator	Area		
	No. 1 highway	No. 2 highway	Conclusions
Pavement	0.15 m bituminous concrete	About 0.15 m bituminous concrete	Basically the same
Cement stabilizing layer	0.4–0.6 m cement stabilizing layer	0.4 m cement stabilizing layer	Basically the same
Subgrade	1.5–2.0 m soil–stone slag mixture (loose subgrade)	1.5–1.7 m lime-treated soil (integral subgrade)	Apparently different
Rheological property of soft soil	11.63 μm/cm rheological settlement	12.01 μm/cm rheological settlement	Basically the same

characteristics conform to the assumptions of the Terzaghi theory, and it requires control over the stability of the state of the subgrade's forced deformation. The result of the existing mechanics analysis of the soft soil structure is less applicable; thus, the differential settlement of the soft soil highway subgrade (including the bumps at bridgeheads) can be resolved when the structural control measure is implemented.

Newtonian mechanics ($F = ma$: structural barycenter is unchanged similar to a mass point) is the source of engineering mechanics. When engineering mechanics ($F = P + T = P_0$) is applied to engineering structural mechanics and deformation issues, its summation process $\Sigma F_i = \Sigma P_{0j}$ shall satisfy two conditions: (1) the stability and environmental adaptability of mass m and movement (deformation) acceleration a of the structure subject to force; (2) the rationality of the transmission force F of the structure subject to force; otherwise, an engineering structural construction control measure is necessary. Engineering mechanics is based on material properties and microstructures. However, when some engineering structures are subject to force, the material properties and microstructures will change, and even the change mode and law are unknown. As a result, the structural constitutive relation, integration and force transferring path will change. When engineering mechanics is applied to engineering structural issues, the engineering structure design and construction satisfies the structural deformation compatibility control conditions. Therefore, (1) the theory of engineering structural stability and balance resolves relatively mature engineering issues, and an accurate analysis method can be employed ($F = P_0$, calculation balance equation); (2) when the engineering structural stability and balance and deformation compatibility control method resolves relatively complicated engineering issues, the engineering structural deformation compatibility control measure can be applied to convert the "leaf issue" to the "apple issue" and then the accurate analysis method can be used to resolve engineering issues ($F = P_0$, calculation balance equation). Obviously, the mechanics analysis method for the soft soil highway subgrade is the second method: apply the deformation compatibility control measure for the subgrade and then conduct mechanics analysis of the soft soil subgrade. In this manner, the error between the calculation results concerning the state of the forced deformation of the soft soil highway subgrade and the actual results can be controlled within the permissible range.

For example, in nature, a single aquatic plant cannot vertically float on a water surface; With natural evolution, the bottom layer of aquatic plants is often intricate and complicated and features a plane structure for stabilizing the entire group of aquatic plants and ensuring that the group of aquatic plants is generally stable and balanced (Fig. 4.4).

A similar philosophy is often applied in civil engineering. For example, a lower diaphragm plate measure is adopted at the soft soil highway subgrade—hand-placed ripraps are paved with concrete, fine-grained soil lime and bonding agents; the lower

Approximate settlement H in soft soil consolidation

(the soil particle structure is maintained, and the settlement is relatively small)

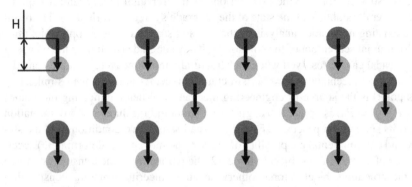

Approximate settlement H in soft soil consolidation and rheology

(the soil particle structure is changed and the settlement is relatively large)

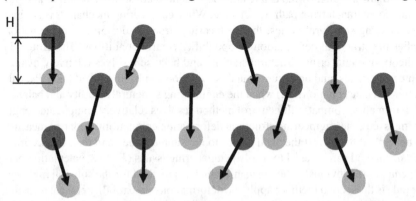

Fig. 4.2 Comparison of the consolidation and rheological coupling settlement of the soft soil highway subgrade

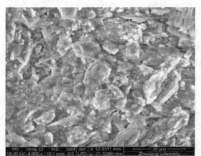
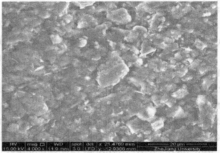

Initial state of soft soil enlarged 4,000 times

State of soft soil compressed at 12.5kp and enlarged 4,000 times

Fig. 4.3 Result of electron microscope scanning involving changes in the microstructure when soft soil is subject to force

Fig. 4.4 Stable and balanced state of a group of aquatic plants

diaphragm plate measure is implemented on high-speed rail subgrade—the bottom is provided with fine-grained soil cement or cement paste, whereas the upper part is paved with fine-grained soil and asphalt. The ultimate goal is to stabilize the state of the subgrade's forced deformation.

Other similar engineering practices are described as follows: (1) when the People's Liberation Army built the Sichuan-Tibet Highway, the army paved wood rafts at the bottom of the marshland subgrade to ensure a stable state of the subgrade's forced deformation; (2) the bottom of the soft soil subgrade in Huangyan, Zhejiang is paved with small wood piles and deep-seated bridge end transition slabs to guarantee a stable state of the subgrade's forced deformation; and (3) the soft soil subgrades of the highways in Wenzhou, Taizhou and Jiaxing, in Zhejiang Province, which have been open to traffic, are consolidated by cement paste; and the subgrade of the connecting line at Huzhou exit, Nanjing-Hangzhou Highway has also undergone consolidation treatment to guarantee a stable forced deformation state.

For the soft soil subgrades with thick soft soil layers and poor soil quality, to effectively promote measures such as the lower diaphragm plates for the soft soil subgrade and other similar engineering measures as well as develop new measures and processes, intuitive judgment can easily derived from heavy parts loaded in car ferries and rubber boats. A hull bottom structure has different rigidity, and heavy parts have different deformation characteristics; deformation differences may be detected among heavy parts in rubber boats, whereas deformation is relatively uniform among heavy parts in car ferries; If the objects loaded in car ferry and rubber boat can control the stability of the forced deformation, the rigidity of the hull bottom structure is not affected. Accordingly, if the stability of the state of the subgrade's forced deformation is controlled or a subgrade baseplate is used to control the stability of the state of the subgrade's forced deformation, the settlement deformation difference in the soft soil subgrade can be effectively controlled.

Overall, when the properties of soft soil, including consolidation, secondary consolidation and rheological properties, are intensively studied, the methods for treating the soft soil subgrade become applicable. When the measures (such as a lower diaphragm plate) for guaranteeing a stable state of the subgrade's forced deformation and the lightweight issue (such as foam concrete) are investigated, the issues of physical concept conformity and accuracy of engineering mechanics analysis can be resolved [1, 2].

References

1. H. Zhu et al., *Engineering Structural Stability, Balance and Deformation Compatibility Control Measures and Applications* (China Communications Press Co., Ltd., Beijing, 2015). (In Chinese)
2. H. Zhu, Z. Zhou, *Thinking of the Issues Concerning Civil Engineering Structure Force Safety* (China Communications Press, Beijing, 2012). (In Chinese)

Chapter 5
Comparative Test, Calculation, and Analysis Concerning the Stability and Looseness of the Soft Soil Highway Subgrade

Given different subgrade structures, the loose subgrade of No. 1 highway and the stable subgrade of No. 2 highway were drilled to obtain cores. Both No. 1 highway and No. 2 highway are located within the same soft soil area and are not located far away from each other.

The soft soil subgrades were vertically sampled at an average depth of 1 m. The comparison of subgrade structures between No. 1 highway and No. 2 highway is shown in Table 5.1 [1–11].

According to the comparison between Fig. 5.1a and b, No. 2 highway adopts a lime-treated soil subgrade that features a stable structure, on which the distribution of forces on the soft soil subgrade is even so that the differential settlement of the subgrade can be controlled. On the contrary, No. 1 highway adopts a soil–stone slag mixture subgrade that features a loose structure, on which the dynamic load

Table 5.1 Comparison of subgrade structures between No. 1 highway and No. 2 highway

Highway	Hole no.	Pavement	Cement stabilized layer	Subgrade
No. 1 highway	HZK2	0–0.15 m asphalt concrete	0.15–0.55 m cement stabilized layer	0.55–2.5 m stone wall (stone and soil mixture)
	HZK3	0–0.15 m asphalt concrete	0.15–0.63 m cement stabilized layer	0.63–2.1 m stone wall (stone and soil mixture)
	HZK4	0–0.15 m asphalt concrete	0.15–0.75 m cement stabilized layer	0.75–2.3 m stone wall (stone and soil mixture)
No. 2 highway	ZK2	0–0.15 m asphalt concrete	0.15–0.52 m cement stabilized layer	0.52–2.2 m lime-treated soil
	ZK3	0–0.15 m asphalt concrete	0.15–0.52 m cement stabilized layer	0.52–2.0 m lime-treated soil
	ZK4	0–0.15 m asphalt concrete	0.15–0.53 m cement stabilized layer	0.52–2.1 m lime-treated soil
Comparison		Basically the same		Apparently different

© Springer Nature Singapore Pte Ltd. and Zhejiang University Press 2019
H. Zhu et al., *Controlling Differential Settlement of Highway Soft Soil Subgrade*,
SpringerBriefs in Applied Sciences and Technology,
https://doi.org/10.1007/978-981-13-0722-5_5

of vehicles makes it easy to slide and rearrange the loose body, which causes local stress concentration and gradual differential settlement of the subgrade.

In the force analysis of the traditional soft soil subgrade, the soil–stone slag mixture subgrade is often considered as a continuous subgrade. Therefore, the vehicle load is considered to be equivalent to a 0.5 m earth-filling static load, which is evenly distributed on the soft soil subgrade. However, this case does not apply to the actual engineering situation. The following specific calculation example is

(a) Soft soil core samples obtained by drilling No.1 highway

(b) Soft soil core samples obtained by drilling No.2 highway

Fig. 5.1 Comparison of core samples of the subgrade soft soil between No. 1 highway and No. 2 highway

considered to quantitatively analyze the forces on the soft soil subgrade under the actual conditions (Fig. 5.2).

According to the *Technical Standard of Highway Engineering*, the following typical vehicle parameters were chosen: vehicle weight of 550 kN (approximately 55 t), four middle wheels and four rear wheels, landing width and length of 0.6 m × 0.2 m, two front wheels, landing width and length of 0.3 m × 0.2 m, and external dimensions of vehicle of 15 m × 2.5 m.

The following values were employed: material density of the soil–stone slag mixture of 2000 kg/m^3; density of the lime-treated soil subgrade of 1730 kg/m^3; and initial rheological stress of the soft soil subgrade of 26 MPa.

The forces on the soft soil subgrade and the subgrade interface in the case of a subgrade with a filled height of 1.5 m are analyzed in Fig. 5.3.

The forces on the soft soil subgrade and the subgrade interface in the case of a subgrade with filled height of 1.2 m are analyzed in Fig. 5.4.

Therefore, the subgrade of No. 2 highway is high, its specific gravity is low (approximately 30%), the total weight of its subgrade is light and its subgrade is integral. Thus, this subgrade satisfies the application conditions of Terzaghi theory, avoids the adverse impact from the concentrated dynamic loads of vehicles and is consistent with the design force model for highway subgrade. This factor is the fundamental factor for controlling the differential settlement of the soft soil highway subgrade within the permissible range.

According to the calculation and analysis results in Figs. 5.3 and 5.4, the vehicle load is considered to be equivalent to the evenly distributed load, and the concentration of the dynamic load of a vehicle is disregarded, which will exert an essential impact on the calculation results, especially for the unstable subgrade. This analysis result corresponded with the actual situation of the differential settlement in highway construction and operation for many years.

According to the comparative test, the calculation and analysis concerning the stability and looseness of the soft soil highway subgrade indicate that both the force properties and deformation properties of the integral subgrade are superior to the force properties and deformation properties of the loose subgrade. Therefore, the design and construction of the highway subgrade should be performed by controlling the

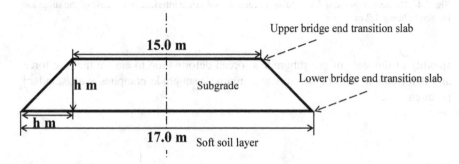

Fig. 5.2 Model for analyzing the forces on the soft soil highway subgrade

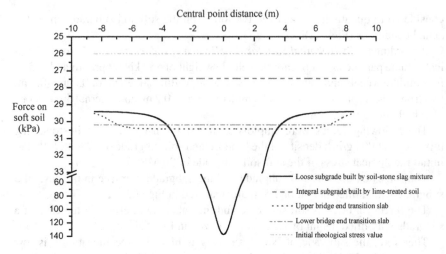

Fig. 5.3 Forces on the soft soil highway subgrade and the subgrade interface in the case of a subgrade with filled height of 1.5 m

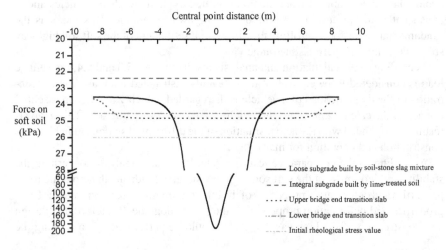

Fig. 5.4 The forces on the soft soil subgrade and the subgrade interface in the case of the subgrade fill height being 1.2 m

stability of the state of the subgrade's forced deformation to ensure that the force and deformation state of the soft soil highway subgrade complies with standard requirements.

References

1. K. Terzaghi, *Theoretical Soil Mechanics* (Wiley, New York, 1943)
2. J. Sun, *The Rheology and Engineering Applications of Geotechnical Materials* (China Architecture & Building Press, Beijing, 1999). (In Chinese)
3. X. Gong, Thinking of numerical analysis of geotechnical engineering. Rock Soil Mech. **32**(2), 321–325 (2011). (In Chinese)
4. J. Sun, *Collected Papers of Academician Sun Jun at the 80th Birthday* (Tongji University Press, Shanghai, 2006). (In Chinese)
5. B. Liu, *Collected Papers of Liu Baochen* (Central South University Press, Changsha, 2011). (In Chinese)
6. H. Zhu et al., *Engineering Structural Stability, Balance and Deformation Compatibility Control Measures and Applications* (China Communications Press Co., Ltd., Beijing, 2015). (In Chinese)
7. H. Zhu, Z. Zhou, *Thinking of the Issues Concerning Civil Engineering Structure Force Safety* (China Communications Press, Beijing, 2012). (In Chinese)
8. Industrial Standard of the People's Republic of China, *JTG D30-2004 Code for Design of Highway Subgrades* (China Communications Press, Beijng, 2004). (In Chinese)
9. D. Gao, *Review and Prospect of Geotechnical Engineering* (China Communications Press, Beijing, 2002). (In Chinese)
10. Y. Guo et al., *New Techniques for Controlling the Post-construction Settlement Deformation in the Subgrades and Treatment of the Bumps at Bridgeheads*, Collected Papers of the 9th Academic Conference on Soil Mechanics and Geotechnical Engineering©, (Tsinghua University Press, Beijing, 2003). (In Chinese)
11. Z. Zhou, *The Theory of Minimum Energy and Its Applications* (Science Press, Beijing, 2012). (In Chinese)

Chapter 6
Comparative Test and Study of the Strength and Rheological Property of the Soft Soil Subgrade

6.1 Comparative Test and Analysis of the Rheological Property of No. 1 and No. 2 Highways' Soft Soil Subgrades

During preparation of the samples of No. 1 and No. 2 highways' soft soil subgrades, great effort has been made to reduce the disturbance caused to soil, to ensure that the mechanical property of indoor soil was consistent with the actual situation, as shown in Fig. 6.1 [1–3].

Ninety-six test soil samples from No. 1 highway and 52 test soil samples from No. 2 highway (148 test samples) were available. The shortest loading time and the longest loading time for each test sample were 144 h and 172 h, respectively. Given the soil quality, the following two loading modes were adopted: 25, 50, 100, and 200 kPa or 50, 100, 200, and 400 kPa. In the initial period of test, data were recorded at the interval of 1 h; in the middle and later periods of test, data were recorded at the interval of 4 h.

Consider No. 5 soil sample (soil depth of 5.8 m) from the No. 2 hole in No. 1 highway as an example, its rheological curve of soft soil is shown in Fig. 6.2. The settlement of soft soil rapidly increased with time at the initial stage of settlement; afterwards, the increase gradually subsided and stabilized. According to vertical observations and comparison of the rheological data for stress levels of 25, 50, 100 and 200 kPa, when the stress level was 25 kPa, the impact of rheology on soft soil settlement can be disregarded. When the stress level was 50, 100 and 200 kPa, rheology exerted a significant impact on soft soil settlement. This finding leads to the concept of "initial rheological stress value"—when the stress level is lower than the initial rheological stress value, the impact of rheology is limited; when the stress level is higher than the initial rheological stress value, ignoring the impact of rheology on the soft soil settlement will exert a non-negligible impact on the results.

To objectively compare the mechanical properties of the soft soil subgrades of No. 1 and No. 2 highways, the impact of the peat soil layer at the soft soil subgrade

© Springer Nature Singapore Pte Ltd. and Zhejiang University Press 2019   31
H. Zhu et al., *Controlling Differential Settlement of Highway Soft Soil Subgrade*,
SpringerBriefs in Applied Sciences and Technology,
https://doi.org/10.1007/978-981-13-0722-5_6

(a) Soft soil core samples obtained by drilling No.1 highway

(b) Soft soil core samples obtained by drilling No.2 highway

Fig. 6.1 Soil samples in the indoor rheological tests of No. 1 and No. 2 highways

of No. 2 highway was excluded and the average loading was calculated. The soft soil layer of No. 1 highway was relatively thin; thus, only the soft soil layers with a depth within 6.5 m were considered for the comparison. The comparison results are listed in Table 6.1.

Table 6.1 Comparison of the mechanical properties of the soft soil subgrades between No. 1 and No. 2 highways

	Rheological settlement value (μm/cm)	Initial rheological stress value (kPa)
No. 1 highway	11.63	26
No. 2 highway	12.01	23

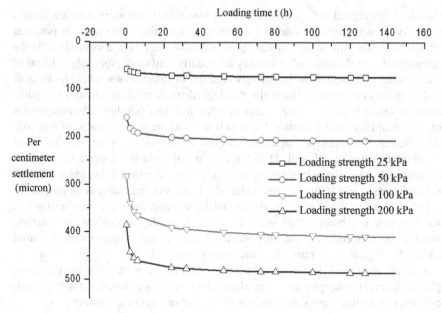

Fig. 6.2 Rheological curve concerning soft soil of No. 1 highway (consider No. 5 sample from No. 2 hole as an example, soil depth of 5.8 m)

According to Table 6.1, the total difference in the engineering properties of the soft soil subgrade between No. 1 highway and No. 2 highway was within 4%; its impact on highway engineering can be disregarded.

Subject to disregarding the impact of the peat soil layer, the final settlement of No. 2 highway's soft soil subgrade was slightly smaller than the final settlement of No. 1 highway's soft soil subgrade. However, the post-construction rheology-caused deformation No. 2 highway's soft soil subgrade will be larger than the post-construction rheology-caused deformation of No. 1 highway's soft soil subgrade. Because No. 2 highway's subgrade was stable, the differential settlement of No. 2 highway's subgrade was smaller than the differential settlement of No. 1 highway's subgrade. Therefore, the study of the differential settlement of the soft soil highway subgrade should focus on not only the mechanical properties of the soft soil subgrade but also its forced deformation mode.

6.2 Comparative Test and Analysis of the Strength and Rheological Property of the Soft Soil Subgrade

From the perspective of mechanics, the initial rheological stress value and the strength value are two completely different concepts. The strength value refers to the maximum stress value that a material can withstand from loading to destruction, whereas the initial rheological stress value refers to the minimum stress value at which distinct rheology occurs where the material is subject to load for a long time. In practical applications, the initial rheological stress value of materials is equivalent to the strength value of the material; thus, they are usually confused, especially in terms of the engineering materials with relatively high rigidity, such as some hard rocks. Soft soil is different from rock; the initial rheological stress value of soft soil is significantly different from the strength value of soft soil. Figure 6.3 shows the comparison of the initial rheological stress value of soft soil and the strength value of soft soil. As indicated, the initial rheological stress value of soft soil is approximately 20% of the strength value of soft soil. Therefore, the initial rheological stress value of the undisturbed soil in the soft soil subgrade is much smaller than the strength value of the undisturbed soil in the soft soil subgrade. If the strength value is employed for calculation and analysis, substantial error will be induced. In particular, when the soft soil subgrade is subject to high stress for a long time, the rheological phenomenon will become especially obvious; the actual calculation and analysis should be based on the rheological value rather than the strength value [1–3].

Overall, the initial rheological stress value and the strength value are two completely different concepts; thus, they should be treated in different ways in the calculation concerning the balance and stability of an engineering structure.

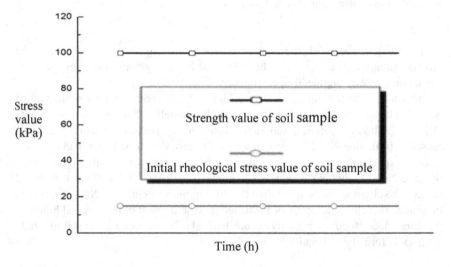

Fig. 6.3 Difference between the strength value of mucky soft soil and the initial rheological stress value of mucky soft soil

References

1. Sun Jun, *The Rheology and Engineering Applications of Geotechnical Materials* (China Architecture & Building Press, Beijing, 1999). (In Chinese)
2. Zhu Hanhua et al., *Engineering Structural Stability, Balance and Deformation Compatibility Control Measures and Applications* (China Communications Press Co., Ltd, Beijing, 2015). (In Chinese)
3. Zhu Hanhua, Zhou Zhihui, *Thinking of the Issues Concerning Civil Engineering Structure Force Safety* (China Communications Press, Beijing, 2012). (In Chinese)

Chapter 7
Basic Characteristics of the Soft Soil Highway Subgrade and Its Design Method

The results of structural comparison between No. 1 highway and No. 2 highway are obtained according to the studies in the previous sections, as shown in Table 7.1 and Fig. 7.1 [1–4].

According to the structural comparison between No. 1 highway and No. 2 highway in Table 7.1 and Fig. 7.1, the largest difference between No. 1 highway and No. 2 highway lies in their subgrades—No. 1 highway adopts the soil–stone slag mixture subgrade which is the loose subgrade; in this case, it is unfavorable for the soft soil subgrade to sustain forces. Conversely, No. 2 highway adopts the lime-treated soil subgrade, which is highly integral; in this case, the concentrated load is evenly distributed to control the deformation of the soft soil subgrade.

To treat the differential settlement of the soft soil highway subgrade, it is necessary to consider the properties of the soft soil subgrade and its forced deformation mode. Two key points should be considered:

(1) Control the stability of the state of the subgrade's forced deformation to evenly distribute the concentrated load, and reduce the peak load and prevent the differential settlement caused by loads that exceed the initial rheological stress value of the soft soil subgrade.

(2) The stability and balance between the subgrade weight and the bearing capacity of the soft soil subgrade are determined by the subgrade weight and the initial rheological stress value of the soft soil subgrade. To be specific: (a) increase the strength of the soft soil subgrade to reduce the rheological settlement of the soft soil subgrade; (b) reduce the upper load of the soft soil subgrade to ensure the subgrade stress is lower than the initial rheological stress value of the soft soil subgrade and the mechanical control goal is achieved. This means the upper load, including the evenly distributed subgrade load and local traffic dynamic load, is lower than the bearing capacity of the soft soil subgrade.

© Springer Nature Singapore Pte Ltd. and Zhejiang University Press 2019
H. Zhu et al., *Controlling Differential Settlement of Highway Soft Soil Subgrade*,
SpringerBriefs in Applied Sciences and Technology,
https://doi.org/10.1007/978-981-13-0722-5_7

Table 7.1 Structural comparison between No. 1 highway and No. 2 highway

	No. 1 highway	No. 2 highway	Conclusion
Pavement	0.15 m bituminous concrete	0.15 m bituminous concrete	Basically the same
Cement stabilizing layer	0.4–0.6 m cement stabilizing layer	0.4 m cement stabilizing layer	Basically the same
Subgrade	1.5–2.0 m soil–stone slag mixture (loose subgrade)	1.5–1.7 m lime-treated soil (stable subgrade)	Apparently different
Rheological property of soft soil	11.63 μm/cm rheological settlement	12.01 μm/cm rheological settlement	Basically the same

(a) Basic situation of stable subgrade of No.2 highway

(b) Basic situation of loose subgrade of No.1 highway

Fig. 7.1 Comparison of the situation of the soft soil subgrades between No. 1 highway and No. 2 highway

According to the above principles, the measures for treating the differential settlement of the soft soil highway subgrade (including the bumps at bridgeheads) can be divided into the following two steps:

(1) Control of the stability of the state of the subgrade's forced deformation: the subgrade can employ EPS, foam concrete, lime-treated soil, lower diaphragm plate, and pile foundation plus sash to control the stability of the state of the subgrade's forced deformation.
(2) Soft soil subgrade design method:

    a. Conduct a rheological test of the soft soil subgrade to determine the initial rheological stress value of the soft soil subgrade.
    b. Calculate the stress value at the bottom of the soft soil subgrade.
    c. Comparison and design: (2.1) if the initial rheological stress value of the soft soil subgrade > the stress value at the bottom of the soft soil subgrade, control the differential settlement of the soft soil subgrade. (2.2) If the initial rheological stress value of the soft soil subgrade is lower than the stress value at the bottom of the soft soil subgrade, employ engineering measures such as a light subgrade and pile foundation plus sash. In this manner, the differential settlement of the soft soil highway subgrade can be controlled within the permissible range.

To resolve the differential settlement of the soft soil subgrade, the forces on the soft soil subgrade must be controlled within the initial rheological stress value of the soft soil subgrade or corresponding engineering measures should be implemented according to the properties of the local soft soil subgrade.

Example 7.1 Based on the design specifications, the soft soil subgrade treatment techniques for treating the bumps at bridgeheads in highways focus on controlling the stability and vertical transition of the state of the subgrade's forced deformation. The vertical transition length of the lower diaphragm plate is associated with the gradient height of the bridgehead subgrade, as shown in Fig. 2.3c. Where the subgrade height ranges from 1 to 1.5 m, only the lower diaphragm plates are adopted. Where the subgrade height exceeds 1.5 m, the lower diaphragm plates and piles for transition (such as the lower diaphragm plate, lime-treated soil, and grouting, in the case of foam concrete and EPS, attention should be paid to the control of the stratification stability and the structural layer on which vehicle loads are evenly distributed) should be adopted.

Example 7.2 As shown in Fig. 7.2, the bridge end transition slab is located at the top of the subgrade; cavity and "seesaw" are easily incurred. The deformation at the bottom of the loose subgrade that is constructed with a soil–stone slag mixture is affected by the rheological and shape distortion properties of the soft soil; thus, a differential settlement occurs. The lines for transferring the additional forces that involve the distribution of the subgrade load differ; generally, the soft soil at the riverside is relatively deep and the subgrade force transfer lines lean to the riverside bridge abutment. Thus, continuous differential settlement may occur in the subgrade or bridge abutment thrust may increase; this finding comprises the analysis of the force and deformation state of the subgrade that involves the bumps at bridgeheads and bridge abutment damage.

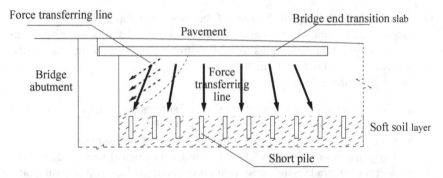

Fig. 7.2 Force and deformation state of the soil–stone slag mixture subgrade after the upper bridge end transition slab treatment at the soft soil subgrade (force transfer path after bridge end transition slab treatment)

Reinforcing bars in deep   EPS light block

Fig. 7.3 Deep bridge end transition slab treatment measure for the soft soil subgrade

Example 7.3 As shown in Fig. 7.3, the bridge end transition slab is located at the bottom of the subgrade, the deformation at the bottom of the loose subgrade made of soil–stone slag mixture is not affected by the rheological and shape distortion properties of the soft soil, and the lines for transferring the additional forces that involve the distribution of subgrade load lean to the riverside bridge abutment. Thus, a continuous even transition of settlement occurs in the subgrade or bridge abutment thrust decreases. This finding comprises the analysis of the force and deformation state of the subgrade, which solves the bumps at bridgeheads and bridge abutment damage. Similar to hold a baby, the forced deformation state of the baby varies with the ways to hold the baby; the people in Yunnan or Guizhou are used to carrying babies on their backs with a haversack to protect the parts of the babies' body that are subject to pressure or bending, including the butt, waist and neck of babies. The state of forced deformation in a baby is generally similar to state of forced deformation in the adult who carries the baby; it is unrelated to the adult's state of motion or the integral displacement of the adult and the baby. Therefore, not only the

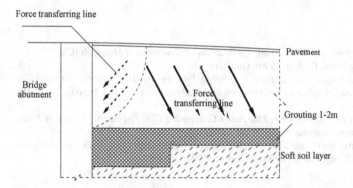

Fig. 7.4 Force and deformation state of the soil–stone slag mixture subgrade after grouting at the bottom of the soft soil subgrade (force transfer path after grouting at the bottom of the subgrade)

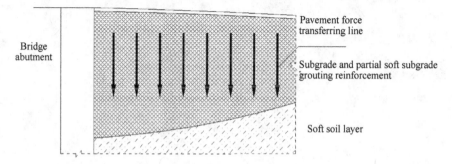

Fig. 7.5 Force and deformation state of the soil–stone slag mixture subgrade after integral grouting in the soft soil subgrade

natural complex state of the forced deformation in a baby but also the ways to hold a baby should be investigated to ensure that the natural complex state of the forced deformation in a baby is converted to a state that is similar to the nature complex state of the forced deformation in the adult who carries the baby. With the help of supporting tools, such as a haversack, both professional personnel and general personnel can carry a baby in a manner that ensures a baby is always in the normal state of forced deformation, which is a simple and reasonable way to carry a baby.

To cope with the bumps at bridgeheads and bridge abutment damage with respect to highways that have been open to traffic, the bridge end transition slabs are placed at the bottom of the subgrade, as shown in Fig. 7.3, or grouting is performed at the bottom of the subgrade, as shown in Fig. 7.4. However, integral grouting in the soft soil subgrade, as shown in Fig. 7.5, entails high costs.

References

1. Sun Jun, *The Rheology and Engineering Applications of Geotechnical Materials* (China Architecture & Building Press, Beijing, 1999). (In Chinese)
2. Zhu Hanhua et al., *Engineering Structural Stability, Balance and Deformation Compatibility Control Measures and Applications* (China Communications Press Co., Ltd, Beijing, 2015). (In Chinese)
3. Zhu Hanhua, Zhou Zhihui, *Thinking of the Issues Concerning Civil Engineering Structure Force Safety* (China Communications Press, Beijing, 2012). (In Chinese)
4. Industrial Standard of the People's Republic of China, *JTG D30-2004 Code for Design of Highay Subgrades* (China Communications Press, Beijng, 2004). (In Chinese)

Chapter 8
Practical Design Method for the Soft Soil Highway Subgrade

The soft soil highway subgrade involves the issue of plane strain, its mechanics characteristics conform to the assumptions of the Terzaghi theory; it requires control of the stability of the state of the soft soil highway subgrade's forced deformation; and the adoption of a synthetic method to cope with the differential settlement of the soft soil highway is necessary. The results of the mechanics analysis of the soft soil structure, especially the results of the deformations, are less applicable. Thus, only when the structural deformation control measure is employed, the differential settlement of the soft soil highway subgrade (including the bumps at bridgeheads) can be resolved [1, 2].

8.1 Design Improvement Techniques for Coping with the Bumps at Bridgeheads of the Soft Soil Highway Subgrade

Significant differential settlement of the soft soil highway subgrade occurs; in particular, the bumps at bridgeheads are more prominent. Currently, many subgrade treatment and analysis methods for the soft soil highway subgrade are available. Why are few successful cases and why are the successful cases not extensively applied? The following two major causes are easily disregarded:

(1) The main difference between a continuous subgrade and a separated subgrade is that the latter can bear the pressure but cannot bear the tensile force and the moment of force, whereas the former can bear the pressure, tensile force and moment of force. **Control the stability of the state of the subgrade's forced deformation is important**.

H. Zhu et al., *Controlling Differential Settlement of Highway Soft Soil Subgrade*, SpringerBriefs in Applied Sciences and Technology, https://doi.org/10.1007/978-981-13-0722-5_8

(2) The soft soil subgrade is characterized by a large void ratio, high compressibility, high moisture content, low permeability, low strength, strong rheological properties, strong structural properties and high sensitivity; it is easy to cause the differential settlement of the subgrade. Thus, **it is very critical to control the unstable continuous settlement caused by the soft soil rheology**.

Existing mechanics analysis and treatment methods for the soft soil highway subgrade assume subgrade deformation compatibility control, which means control the stability of the state of the subgrade's forced deformation, bearing pressure, pulling and moment of force. However, a practical operation often does not take controlling over the stability of the subgrade's forced deformation state into consideration. For example, the soil–stone slag mixture subgrade is only able to bear pressure but fails to withstand the differential settlement. Engineering construction personnel generally consider the impact of the consolidation and secondary consolidation of the soft soil subgrade. However, they neglect the control of the unstable continuous settlement caused by the soft soil rheology. Therefore, significant differential settlement of the soft soil rheology is induced; particularly, the bumps at bridgeheads are more prominent.

Existing analysis and treatment methods for the soft soil subgrade have two problems: the traditional soft soil subgrade analysis theory adopts deterministic methods, such as a spring, dashpot and sliding block, to simulate the uncertain movement characteristics of the soft soil subgrade (particles), and internationally, it is widely believed that force calculations concerning the soft soil subgrade are relatively accurate with large errors in the deformation calculation, which conflicts with the functional relationship $y = f(x)$. Theoretical logic is rigorous, but does not conform to dialectical thinking; only when regular movements of the soft soil particles are technically controlled, which means the stability of the subgrade is simply controlled, the results of the soft soil subgrade deformation calculation can then fall within the permissible engineering range.

Based on experience summarization and dialectical thinking, the design of the soft soil subgrade is improved by controlling the stability of the state of the soft soil subgrade's forced deformation and the unstable continuous settlement caused by the soft soil rheology. Consider a 1.5 m-high subgrade as an example, as shown in Chap. 5.

According to Figs. 5.3 and 5.4, soft soil highway subgrade design methods are improved according to the structural deformation compatibility control method. Based on the design specifications, design is conducted by controlling the stability of the state of the subgrade's forced deformation and its vertical transition, whereas the key technique consists of placing baseplates or sashes and vertical transition pile foundations for controlling the differential settlement of the subgrade—the vertical transition length of the lower diaphragm plates and piles are related to the gradient height of bridgehead subgrade. When the subgrade height ranges from 1–1.5 m, only the lower diaphragm plates are adopted. When the subgrade height exceeds 1.5 m, the lower diaphragm plates and piles, such as the lower diaphragm plate, lime-treated soil, grouting, must be adopted for transition. In the case of foam concrete, and EPS, attention should be paid to control of the stratification stability and the structural layer on which vehicle loads are evenly distributed. To control the differential settlement of the bridgehead subgrade within the permissible engineering range, and effectively cope with the bumps at the bridgeheads on the soft soil highway subgrade.

Different methods can be adopted to treat the newly built soft soil highway subgrades and the existing soft soil highway subgrades.

(1) Methods for treating the newly built soft soil highway subgrades

(1) According to the comparison of the subgrade weight per unit area and the initial rheological value of the soft soil subgrade, if the subgrade weight per unit area is larger than the initial rheological value of the soft soil subgrade, the addition of subgrade baseboards or sashes and short piles or a vertical transition pile foundation, as shown in Figs. 2.1 and 2.3c, is necessary. In this manner, the stability of the state of the soft soil subgrade's forced deformation and the unstable continuous settlement caused by the soft soil rheology can be controlled. If the subgrade weight per unit area is smaller than the initial rheological value of the soft soil subgrade, only the addition of subgrade baseboards or sashes is necessary and short piles or a vertical transition pile foundation is needed.

(2) The surface of the soft soil subgrade is manually leveled to maintain a vertical length of 6–20 m and tilting of 3°–5° in the direction of the subgrade at the outer side of the bridge. When heavy-duty machines are used to perform the leveling of the soft soil subgrade's surface, the hard shell of the soft soil subgrade's surface can be easily destroyed. Thus, special attention should be paid to this issue at the time of leveling to prevent backfire.

(3) The manually leveled surface of the soft soil subgrade is paved with 30–50 cm-thick sand grains or broken stone hardcore as the mechanical cushion layer for the next operation and the drainage layer on the surface of the soft soil subgrade.

(4) In response to the requirements in Step (1), short piles in two rows are horizontally placed in the soft soil subgrade along the bridge edge, or the vertical transition pile foundations in one row are vertically placed in the soft soil subgrade at the interval of 10 m starting from the bridge edge. These piles or pile foundations provide support for the subgrade baseboards or sashes to control the unstable continuous settlement caused by the soft soil rheology.

(5) The reinforced concrete baseplates or sashes are cast in place or prefabricated reinforced concrete baseplates or sashes are placed on the existing 30–50 cm-thick sand grain layer or broken stone hardcore within the vertical 6–20 m-long range of the soft soil subgrade's surface. These baseplates or sashes provide a layer for controlling the stability of the state of the soft soil subgrade's forced deformation and the unstable continuous settlement caused by the soft soil rheology. Consequently, **the bumps at bridgeheads of the soft soil highway subgrade are resolved**.

(6) The subgrade and pavement are built according to the specifications.

(2) Methods for treating the existing soft soil highway subgrade

(1) The existing pavement of the entire subgrade is removed for 6–12 m on a
layered basis from the bridge edge but the highway shoulder is retained to
restrict the lateral deformation of the subsequent light materials. The thickness of the removed subgrade pavement can be determined to the extent
that the weight of the removed existing single-lane subgrade pavement is
larger than the sum of the weight of the substituted light materials and the
load of one heavy-duty vehicle, and the safety factor ranges from 1.2–1.5.

(2) The bottom layer of the removed subgrade is leveled, the reinforced concrete baseplates or sashes are cast in place or prefabricated reinforced
concrete baseplates or sashes are placed, as the layer for controlling even
forced deformation of the subsequent light materials.

(3) On the cast-in-place reinforced concrete baseplates or sashes or the placed
prefabricated reinforced concrete baseplates or sashes, the light material
structural layer is cast in place or the prefabricated light material structural
layer is placed to mitigate the subgrade weight so as to control the unstable
continuous settlement caused by soft soil rheology.

(4) The structural layer for pavement enhancement is placed to average the
loads of heavy-duty vehicles and jointly control the stability of the state
of the soft soil subgrade's forced deformation and the unstable continuous
settlement caused by the soft soil rheology, **which effectively resolve the
bumps at the bridgeheads of the soft soil highway subgrade**.

8.2   Techniques for Treating the Bumps at the Bridgeheads of the Operating Highway Based on Structural Deformation Compatibility Control

A bump is formed at a highway bridgehead when vehicles that rapidly cross the
boundary between the bridge and the highway bump jump as a result of the differential settlement or large longitudinal slope between the structures in operating
highway projects, such as bridges and culverts, and the embankment at the abutment
back, as shown in Fig. 8.1. Currently, the bumps at bridgeheads are generally treated
by grouting to reinforce the soft soil subgrade and adding abutment end transition
slabs in the upper part. However, the key issues that affect the bumps at bridgeheads are not specified. As shown by the coring results from numerous sites where
grouting was conducted to resolve the bumps at the bridgeheads of the soft soil subgrades, basically there are no specific reasonable structural bearing systems, only tree
root-like grout veins are available in the middle of subgrades or soft soil subgrades
[1, 2].

Techniques for treating the bumps at bridgeheads of operating highways do not
explicitly reveal indicators such as the stability of the state of the subgrade's forced
deformation and soft soil rheology. This link is the key link that affects bumps at

Fig. 8.1 Bump at highway bridgehead

bridgeheads. Thus, an effective treatment technique consists of placing subgrade bearing piles and conducting high-pressure grouting at the bottom of an embankment to form a reinforcing layer with a certain thickness, as shown in Fig. 8.2. The embankment grouting reinforcing layer has a certain strength and rigidity. Thus, the upper embankment and vehicle load can be transferred to the lower part, and the embankment grouting reinforcing layer, piles, and soft soil jointly contribute to from a subgrade bearing system, so as to stabilize the state of the subgrade's forced deformation, effectively control the subsequent settlement of the subgrade and resolve the bumps at bridgeheads. The force diagram of the bearing system in this process is shown in Fig. 8.3.

The construction process for treating the bumps at the bridgeheads of an operating highway is described as follows:

(1) Removal of the original pavement

The original pavement can be removed by a pavement milling machine or backhoe excavator vibration head that is consistent with the pavement type. After the pavement is removed and slag is cleared, construction of subgrade bearing piles (cement-mixed piles) begins.

(2) Construction of subgrade bearing piles (cement-mixed piles)

    (1) The bearing piles are soil–cement-mixed piles; construction is performed by single-axis bidirectional soil–cement-mixed pile machines.
    (2) Arrangement of the bearing piles:
        a. Pile diameter of 650 mm
        b. They feature a quincunx arrangement at a plane; the vertical and horizontal central distance between two piles is 1000 mm.
        c. Nine rows are vertically arranged. The edge of the bearing piles in the first row is 500 mm away from the edge of the bridge abutment; the

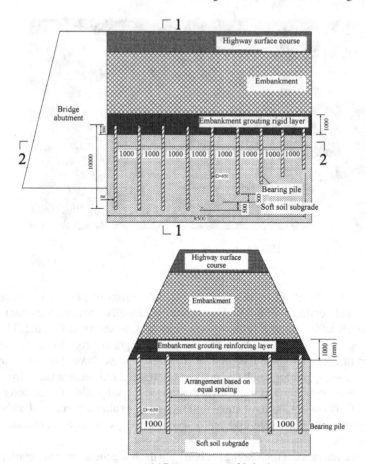

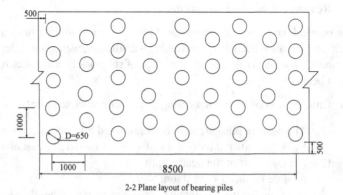

1-1 Transverse structure of the bearing system

2-2 Plane layout of bearing piles

Fig. 8.2 Vertical construction of the bearing system (unit: mm)

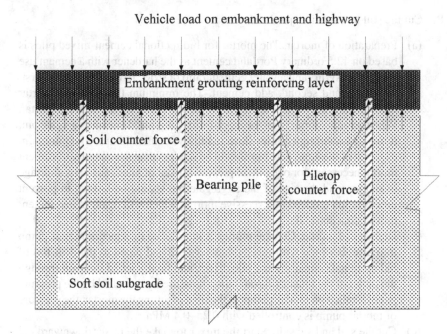

Fig. 8.3 Force diagram of the bearing system

total length from the edge of the bridge abutment to the central line of the piles in the last row is 8500 mm.

   d. The vertical length of the piles in the first to fourth rows is 10,000 mm. The pile length is reduced to 500 m on a row-by-row basis from the piles in the fifth row.

   e. The bearing piles are available at a unified length in the horizontal direction along an embankment. The number of horizontal bearing piles is determined according to a spacing of 1000 mm and the subgrade width.

   f. The bearing piles are buried, for 300 mm, in the upper embankment grouting reinforcing layer.

(3) Construction of soil–cement-mixed piles:

   A. The mixed piles and pile foundation are installed in place.

     (a) Conduct onsite surveying and positioning according to the pile location arrangement diagram. Set the accurate pile locations on the embankment, and place a small wood pile at each pile location to ensure that the pile driver can be aligned with reasonable pile locations.

     (b) Install the pile foundation and ensure centering

B. Cut the soil and subside, jet the mortar

   (a) Preparation of mortar. The mortar for bidirectional cement-mixed piles is based on 42.5 ordinary Portland cement as the hardener with a cement use level of 65 kg/m^3 and a water cement ratio of 0.50. To mix the mortar, first, add water and second, add cement. The minimum time of each mortar mixing shall not be less than 2 min for fully mixing the cement mortar. When cement mortar from the mortar mixer is poured into an aggregate bin, it must pass through a filter sieve to remove cement lumps. The minimum capacity of the aggregate bin should be not less than the quantity of cement mortar necessary for one pile to prevent broken of piles caused by shortage of mortar, and the size of the aggregate bin should be limited to prevent precipitation induced by excessive mortar, which will result in insufficient mortar concentration.

   (b) Delivery of mortar. Cement mortar is introduced by squeeze mortar pump into a rubber pipe with inner diameter Ø32 and piped to the drill rod of bidirectional cement-mixed pile driver, and finally jetted into soil through the outlet of the inner drill rod; pump pressure is controlled by the pressure gage installed on mortar pump; at the time of jetting the mortar, the pressure of mortar pump is controlled with 0.25–0.4 MPa.

   (c) Cut the soil and subside. Start the mixer to make the mixer downward cut the soil along the guide frame, meanwhile start mortar pump to jet cement mortar into the soil. Two groups of blades rotate forward and reversely, cut and stir the soil to the design depth. The mixer jets and stirs mortar at the pile bottom for at least 15 s. The minimum rotating speeds of the inner and outer drill rods at the time of drilling is 50 r/min; the drilling speed of the drill rods is controlled within 0.5–0.8 m/min.

C. Lift the mixer

   (a) Close the mortar pump and lift the mixer; the two groups of blades rotate forward and reversely to stir the cement soil until the upper surface of the embankment.

   (b) When lifting, it is necessary to strictly control the lifting speed and stir the cement soil into which the mortar has been jetted; the rotating speed of the inner and outer drill rods at the during lifting is controlled to be not less than 50 r/min; the lifting speeds of the drill rods is controlled within 0.7–1.0 m/min.

D. Shift the pile driver to the next pile location

(3) Construction of the embankment grouting reinforcing layer

   (1) Use the single-pipe high-pressure jet grouting method to consolidate the fillers within the scope of 1 m from the embankment bottom, which is to form a 1 m-thick embankment reinforcing layer at the embankment bottom, with an aim to evenly transferring the upper load to the bearing piles and the soft soil subgrade.

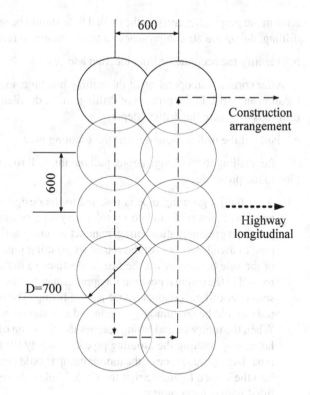

Fig. 8.4 Arrangement and construction of jet grouting piles (unit: mm)

(2) Ensure that the 700 mm-diameter single-pipe high-pressure jetted grouting piles with a 600 mm spacing between centers engage with each other to form a stable embankment reinforcing layer; the thickness of the reinforcing layer is 1 m. The arrangement and construction of jet grouting piles is shown in Fig. 8.4.

(3) The high-pressure jet grouting construction within the scope of pavement is performed at the subgrade top, whereas the high-pressure jet grouting construction beyond the scope of pavement is performed by the construction platform near the subgrade slope.

(4) The main machines in the single-pipe high-pressure jet grouting construction include jumbolter MGJ-50, mortar pump XPB-90KW, and mortar mixer 2.2 kW.

(5) The main construction process of single-pipe high-pressure jet grouting is described as follows:

   a. Position the drilling machine

Shift and place the pile driver according to the design pile driver construction procedure and pile locations. After the drilling machine is positioned, level and center the pile foundation and adjust the perpendicularity of the pile driver to ensure that drill rod is aligned with the pile location. The error should be within 10 mm; the error in the perpendicularity of the drilled hole should be smaller than 0.3%. Prior to drilling, debug the air compressor and mortar pump to facilitate normal operation.

b. Identify the accurate drilling location and drill

After normal trial operation of the drilling machine, identify the accurate drilling location and drill. In the process of drilling, make detailed records of the number of drill rods to ensure the drilling depth.

c. Pull out the drill rods, and insert the grouting pipe

After drilling to the design depth, pull out the drill rod, and insert the jet grouting pipe to the preset depth.

d. After the jet grouting pipe is inserted to the design depth, connect the mortar pump and then perform bottom-top rotary jet grouting. When jetting, attain the preset jet pressure; after jetting, conduct grouting and lift the rotary jet grouting pipe to avoid twisting off the rotary jet grouting pipe. To guarantee the quality of the pile bottom, when the nozzle declines to the design depth, it should be rotated in the original position for approximately 10 s. After the mortar normally spill over orifice, conduct rotary jet and lifting. The rotation and lifting of drill rods should be continuously conducted and should not be interrupted.

e. When the rotary jet and lifting reaches the elevation of the top of the reinforcing layer, stop rotating the grouting pipe and slowly lift the orifice of the grouting pipe. During this process, the mortar pump should continue injecting the mortar into the orifice to ensure that the drilled holes above the reinforcing layer are filled with cement mortar.

f. Control the parameters of the high-pressure rotary jet: water cement ratio in cement mortar: 1:1 (42.5 ordinary Portland cement is adopted). The grouting pressure: higher than 20 MPa. The jet grouting pipe lifting speed: 20–28 cm/min. The rotating speed of the jet grouting pipe: 20–25 r/min. Cement use level: not less than 210 kg/m^3.

(4) Embankment restoration and pavement construction

(1) Treatment of disturbance in the embankment filling body

During grouting construction in the bearing piles and rigid layer, the subsidence and lifting of the mixed piles will cause stirring disturbance to the embankment within the scope of the pile diameter and destroy the original density of the embankment fillers. During grouting construction in the rigid layer, the drilling machinery will also cause disturbance to the embankment fillers, and leave drilled holes. Therefore, it is necessary to treat the disturbance in the embankment during construction.

a. During the construction of the cement soil mixed piles, in the part of the embankment above the elevation of the top of the bearing pile, only stirring is conducted and no grouting is performed during the subsidence and lifting of the mixer.

b. During grouting construction in the rigid layer, grout holes are designed at the locations of the mixed piles. After grouting at the rigid layer, grouting reinforcement is installed at the part of the embankment that was disturbed by the mixed pile driver above the rigid layer, and grouting is conducted according to the grouting requirements of the rigid layer.

c. After grouting in the rigid layer is completed, during the process of lifting the grouting pipe, pressureless grouting is conducted in the drilled holes that are formed by high-pressure grouting at the part of the embankment above the rigid layer. Thus, the drilled holes are filled with cement mortar, which becomes the cement filling body after setting.

With the above treatment method, the embankment structure is generally reinforced while the disturbed part of the embankment is reliably treated.

(2) Construction of highway surface

Perform the highway surface construction according to design requirements.

8.3 Methods for Controlling the Differential Settlement of the Soft Soil Highway Subgrade

Differential settlement is relatively serious in existing soft soil highway subgrades. In particular, horizontal and vertical cracking occurs on the pavement of the soil and stone-filled subgrades. However, horizontal and vertical cracking seldom occurs on the pavement of the stable subgrades, such as lime-treated soil subgrades and soil and stone subgrades, which are subsequently grouted. The core issue in the differential settlement of a soft soil highway subgrade is to control the stability of the subgrade's forced deformation state. If actions are taken to economically stabilize the state of forced deformation in the newly built highway subgrades in the coastal areas with numerous mountains, minimal land and abundant soil–stone slag mixtures, repeated horizontal and vertical cracking in the pavement can be prevented, which saves construction and maintenance costs [1, 2].

Given that the stability of the state of forced deformation is not effectively controlled in the newly built soil–stone slag mixture subgrades, the solution is to implement measures as a frame between the bottom layer of the subgrade and the top layer of the soft soil subgrade to stabilize the state of forced deformation in the newly built soil–stone slag mixture subgrades, and prevent horizontal and vertical cracking in the pavement.

The methods for controlling the differential settlement of the soft soil highway subgrade are described as follows:

(1) According to the properties of the soft soil and relevant specification requirements, $\varphi 50$ cm cement-fixed piles are placed with a length range from 0.0–8.0 m (the length of the cement-mixed piles is 0 m, and the reinforced concrete frames are directly placed into the hard shell layer of the soft soil subgrade), and plane arrangement features with a 2.0 m × 2.0 m rectangular array, as shown in Figs. 8.5 and 8.6.

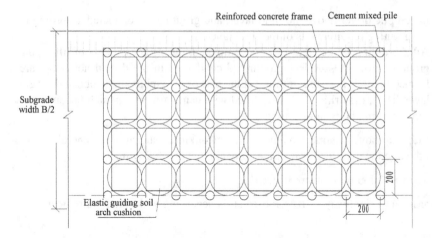

Fig. 8.5 Plane arrangement of the soft soil highway subgrade (unit: cm)

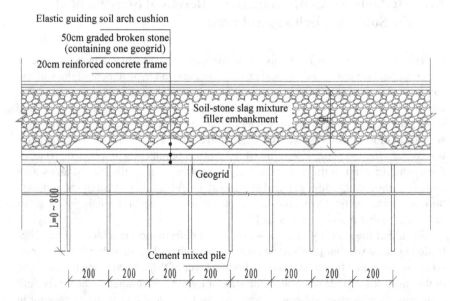

Fig. 8.6 Vertical section arrangement of the soft soil highway subgrade (unit: cm)

(2) The subgrade is leveled according to the subgrade height and relevant specification requirements; according to the plane arrangement requirements for
highways, soil grooves for the sashes, which feature a rectangular array, are
excavated in the direction of the subgrade. The depth and width of the soil
groves are 20 and 50 cm, respectively; 20 cm high and 50 cm wide sash beams
are cast in place in soil grooves and maintained for 24 d; The sash size in the
reinforced concrete frame is 2.0 m × 2.0 m. Ultimately, Reinforced concrete

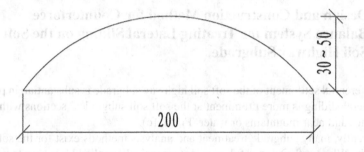

Fig. 8.7 Vertical view of the elastic guiding soil arch cushion of the soft soil highway subgrade (unit: cm)

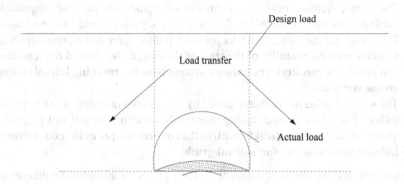

Fig. 8.8 Load transfer on the elastic guiding soil arch cushion of the soft soil highway subgrade

frames (the length of each frame is 20 m long) are constructed; the sashes correspond to cement-fixed piles and the upper part of the apex of each sash is provided with cement-fixed piles.

(3) One geogrid is placed on the reinforced concrete frames.
(4) One elastic guiding soil arch cushion, which is similar to a part of a sphere, is placed on the geogrid on each reinforced concrete frame. This cushion is fixed with a geogrid, as shown in Figs. 8.7 and 8.8. The basal diameter of the elastic guiding soil arch cushion is 2 m, its height ranges from 0.3–0.5 m, and its material is hard foam. The elastic guiding soil arch cushion and the center of the frame share the same vertical line, and the basal circle of the elastic guiding soil arch cushion is tangent to four beam center lines of the frame. Thus, the basal rim of the elastic guiding soil arch cushion is supported by a sash beam.
(5) 30–50 cm graded broken stones are placed on the geogrid and the elastic guiding soil arch cushion
(6) According to the specification requirements, one layer is available at an interval of 50 cm, and the soil–stone slag mixture subgrade is built on a layered basis until the design height is attained.

8.4 Design and Construction Method for Counterforce Balance System for Treating Lateral Sliding on the Soft Soil Highway Subgrade

The differential settlement of the soft soil highway subgrade is substantial; in particular, lateral sliding is more prominent at the soft soil subgrades' sections with poor properties and near mountains or water (Fig. 1.1c).

Currently, many subgrade treatment and analysis methods exist for the soft soil subgrade. Why are the successful cases not extensively applied? It is easy to neglect the following two major causes:

(1) The main difference between a continuous subgrade and discrete subgrade is the latter can bear pressure but fails to bear the pulling force and moment of force, whereas the former can bear pressure, pulling force and moment of force. **Control over the stability of the state of the subgrade's forced deformation can provide a counterforce balance system base for resisting lateral sliding on the subgrade**.

(2) The soft soil subgrade is characterized by a large void ratio and high compressibility of soft soil. Lateral sliding may easily occur on the soft soil subgrade. Therefore, **it is very critical to control the counter torque in the counterforce balance system of the soft soil subgrade**.

The traditional soft soil subgrade analysis theory does not take consider the control of the counterforce balance system of the soft soil subgrade. When the stability of the state of the soft soil subgrade's forced deformation is simply controlled and a counter torque is provided, lateral sliding on the soft soil subgrade can be prevented. Based on experience summarization and dialectical thinking, the soft soil subgrade design is improved by controlling the stability of the state of the soft soil subgrade's forced deformation and providing a counter torque [1, 2].

As shown in Fig. 8.9, the balance system and counterforce balance system equations for the integral subgrades or the subgrades with baseplates are expressed as

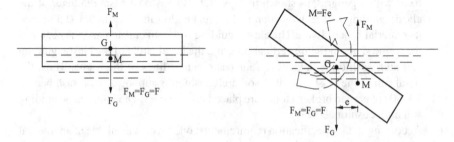

Fig. 8.9 Force analysis of measures for controlling the counter torque in the counterforce balance system of the soft soil subgrade

$$F_G = F_M \tag{8.1}$$

$$M_F = F_M * e \tag{8.2}$$

where F_G is the subgrade weight; F_M is the subgrade counter force; M_F is the subgrade counter torque; and e is the theoretical counter torque of the subgrade counterforce balance system.

Equations (8.1) and (8.2) and Fig. 8.9 show that the soft soil subgrade design method is improved according to the structural deformation compatibility control method. The design controls the stability of the state of the subgrade's forced deformation and its lateral resistance. Its key technique consists of arranging the baseplates or sashes for controlling the differential settlement of the subgrade and lateral resistance piles and building the counterforce balance system for controlling the soft soil subgrade to control the lateral sliding on the soft soil highway subgrade.

Therefore, to prevent and control lateral sliding on the soft soil highway subgrade, the following different construction methods for the newly built soft soil subgrades near mountains and water can be adopted

(1) Newly built soft soil subgrades near mountains

(1) The scope in which the lateral resistance piles are arranged for the soft soil subgrades near mountains is qualitatively analyzed according to Eq. (8.2). Based on the soft soil subgrade design specification, the subgrade baseplates or sashes are arranged, the lateral resistance piles are placed, and **the counterforce balance system for resisting the lateral sliding on the soft soil highway subgrade is provided**.

(2) The surface of the soft soil subgrade is manually leveled. When heavy-duty machines are used to perform the leveling of the soft soil subgrade's surface, it is easy to destroy the hard shell of the soft soil subgrade's surface, thus special attention should be paid to this matter at the time of leveling or it will backfire.

(3) The manually leveled surface of the soft soil subgrade is paved with 30–50 cm-thick sand grains or broken stone hardcore as the mechanical cushion layer for the next operation and the drainage layer on the surface of the soft soil subgrade.

(4) In response to the requirements in Step (1), the soft soil subgrade is provided with the lateral resistance piles according to the design requirements, which generally consists of Ø100 cm reinforced concrete piles and features a horizontal and vertical 2×10 m rectangular arrangement. Piles penetrate the soft soil layer to enter the supporting layer, which can reinforce the soft soil subgrade and provide the passive resistance by the counter torque (Fig. 8.10).

(5) Based on the design requirements, the reinforced concrete baseplates or sashes are cast in place or prefabricated reinforced concrete baseplates

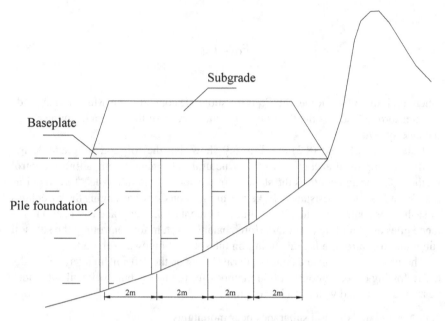

Fig. 8.10 Construction of the baseplate or sash for controlling the lateral sliding on the soft soil subgrades near mountains and the lateral resistance piles

or sashes are placed on the existing 30–50 cm-thick sand grain layer or broken stone hardcore on the surface of the soft soil subgrade, as the layer for controlling the stability of the state of the soft soil subgrade's forced deformation. They are combined with the pile foundation to shape **the counterforce balance system for resisting the lateral sliding on the soft soil highway subgrade and controlling the lateral sliding on the soft soil highway subgrade**.

(6) The subgrade and pavement are built according to the specifications.

(2) The newly built soft soil subgrades near water

   (1) The scope in which the lateral resistance piles are arranged for the soft soil subgrades near water is qualitatively analyzed according to Eq. (8.2), and then based on the soft soil subgrade design specification, the subgrade baseplates or sashes are arranged and the lateral resistance piles are placed, and **the counterforce balance system for resisting the lateral sliding on the soft soil highway subgrade is provided**.

   (2) The surface of the soft soil subgrade is manually leveled. Special attention should be paid to preventing damage to the hard shell on the surface of the soft soil subgrade at the time of leveling.

   (3) The manually leveled surface of the soft soil subgrade is paved with 30–50 cm-thick sand grains or broken stone hardcore as the mechanical

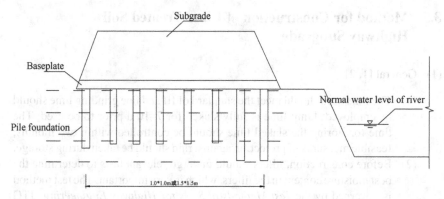

Fig. 8.11 Construction of the baseplate or sash for controlling the lateral sliding on the soft soil subgrades near water and the lateral resistance piles

cushion layer for the next operation and the drainage layer on the surface of the soft soil subgrade.

(4) In response to the requirement in Step (1), the soft soil subgrade is provided with the lateral resistance piles according the design requirements, which are generally Ø16 cm single-walled threaded corrugated pipes filled with foam concrete. Its process is described as follows: first, use follow-up steel pipes to press the corrugated pipes, and second add foam concrete. This design features a 1.0 × 1.0 m or 1.5 × 1.5 m quincunx arrangement; the length of the pile foundation ranges from 6–10 m (Fig. 8.11). The foam concrete mixing ratio is expressed as follows: foam particles:sand:cement:water:admixture = 14.6:562:280:146:6.0 (its unit weight = water unit weight 10 kN/m^3). This ratio can reinforce the soft soil subgrade and provide the active resistance, which produces counter torque.

(5) Based on the design requirements, the reinforced concrete baseplates or sashes are cast in place or prefabricated reinforced concrete baseplates or sashes are placed on the existing 30–50 cm-thick sand grain layer or broken stone hardcore on the surface of the soft soil subgrade, as the layer for controlling the stability of the state of the soft soil subgrade's forced deformation. They are combined with the pile foundation to shape **the counterforce balance system for resisting the lateral sliding on the soft soil highway subgrade and control the lateral sliding on the soft soil highway subgrade**.

(6) The subgrade and pavement are built according to the specifications.

8.5 Method for Construction of Lime-Treated Soil Highway Subgrade

(1) General [1, 2]

(1) The lime used should meet the standard of III and fine grinding lime should be employed; lump lime is fully slaked for 7–10 d prior to be used; The time for storing the slaked lime should be controlled within two months; feasible measures of protection against rain should be taken during storage.

(2) Before construction, fillers should undergo relevant tests to determine the best moisture content in the fillers, which is very important. The test method is governed by the *Test Methods of Soils for Highway Engineering* (JTG E40-2007). In the compaction test on mixture, the lime should be the same as that used at the construction site.

(3) The amount of lime and soil used should be accurately controlled according to the design requirements, and evenly stirred. The stirred depth should reach the bottom layer of the road with the unslaked lime should be removed.

(4) Where the moisture content in the filling undisturbed soil is excessively high or the undisturbed soil is cohesive soil, the secondary lime mixing process should be considered, (the first time) 2–7% quick lime is added to the borrow pit, and (the second time) the slaked lime is added to road mixing. When the moisture content in the undisturbed soil falls within +5 to +3% of the best moisture content, the construction process of primary lime mixing—road mixing—should be adopted.

(2) Construction process

The construction process flow for the lime-treated soil subgrade is shown in Fig. 8.12.

(3) Key points of construction

(1) The primary or secondary lime mixing process is adopted based on the moisture content in the soil or the soil quality. The filling soil for the primary lime mixing process can be directly obtained from a borrow pit. Where the secondary lime mixing process is adopted, an excavator is used to excavate the lime-treated soil in a borrow pit to be stacked for slaking for 3 d. In the process of slaking in the borrow pit, it is necessary to daily turn and stir the slaking material and the lime-treated soil in borrow pit should be stirred evenly.

(2) The spacing for stacking up the soil is calculated according to the volume of a dumper. Grids are built at the lower bearing layer. The plain soil from a borrow pit or the soil that is "sanded" and treated by lime for the first time is dumped. A bulldozer is used to evenly spread the soil according to the lay-down thickness.

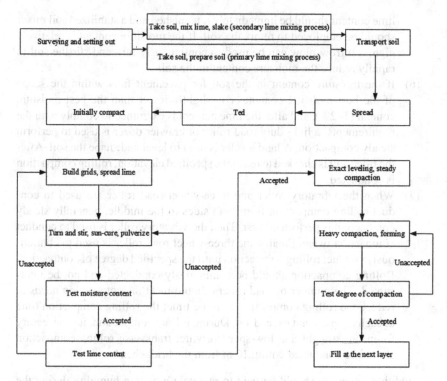

Fig. 8.12 Construction process flow chart for the lime-treated soil subgrade

(3) A furrow plow is used for continuous turning and stirring to reduce the moisture content in soil. When the moisture content in the soil spread on the site decreases to +3 to +5% of the best moisture content, use light-duty road roller to flatten the soil is appropriate. Grids are built on the ground. The grid area is controlled to the extent that each grid can accommodate the slaked lime from one vehicle, and then lime is spread by bulldozer or manually spread.

(4) The stabilized soil mixer is used to mix and pulverize the lime-treated soil for at least two times. Mixing should be performed deep into the lower bearing layer for 5–10 mm, with no plain soil interlayer allowed. Sieving is conducted to test the size of the soil particles on the site. The control standard is described as follows: the content of particles larger than 5 cm should be less than 5%. The content of particles larger than 2 cm should be less than 20%; for the soil with much concretion, the content of particles larger than 2 cm should be less than 30%.

(5) After soil blocks are pulverized, the lime content and moisture content should be tested in a timely manner. The test standard for lime content is defined as follows: the lime content at all test points should be greater than the design lime content (−1%). If the lime content is relatively low, the

lime content should be immediately replenished and a stabilized soil mixer should be employed to remix the soil. If the moisture content is relatively high, a furrow plow can be used to continuously turn and stir the soil to rapidly reduce the moisture content in the soil.

(6) If the moisture content in the soil for pavement falls within the scope of the best moisture content or is slightly lower than the best moisture content (1–2%), and after the lime content is examined to comply with the requirements, a light-duty road roller or crawler dozer is used to perform steady compaction. A land leveler is used to level and shape the soil. After the filled soil is checked to reach the specified elevation, rolling compaction is conducted.

(7) When the vibratory roller and three-wheel road roller are used to conduct rolling compaction from both sides to the middle, generally steady compaction is performed first. Then the vibratory roller is used to conduct compaction twice. Finally, the three-wheel road roller is used to continuously conduct rolling compaction until the specified degree of compaction. Rolling compaction should be continuously conducted and not be interrupted. The number of road rollers should meet the compaction needs to reduce the rolling compaction forming time; the rolling compaction time at each layer cannot exceed 1 d. During rolling compaction, it is necessary to ensure straight and low-speed driving; transverse rolling compaction should be conducted within 10 m from the bridgehead.

In addition, attention should be paid to maintain a certain humidity during the curing period; excessive moist or quick switching between dryness and wetness is not allowed.

References

1. Zhu Hanhua et al., *Engineering Structural Stability, Balance and Deformation Compatibility Control Measures and Applications* (China Communications Press Co., Ltd, Beijing, 2015). (In Chinese)
2. Zhu Hanhua, Zhou Zhihui, *Thinking of the Issues Concerning Civil Engineering Structure Force Safety* (China Communications Press, Beijing, 2012). (In Chinese)

Part II
Practice in the Improved Design Engineering Construction of the Soft Soil Highway Subgrade

Chapter 9
Practice in the Construction of the Lime-Treated Soil Highway Subgrades in Jiaxing City

9.1 Main Differences in Application Between Soil–Stone Slag Mixture Subgrade and Lime-Treated Soil Subgrade

(1) The soil–stone slag mixture construction control has difficulty in satisfying the requirements, which may easily cause differential settlement in highway during service. The lime-treated soil particles are uniform, and a plate body structure can form after construction.

(2) The water stability of the soil–stone slag mixture is low, water can easily permeate the subgrade, which induces an adverse impact. The water stability of the lime-treated soil is high; the resulting plate body structure exhibits a certain water-resisting property.

(3) The exploitation of the soil–stone slag mixture causes severe damage to the local ecological environment. From 2013, the government shut down numerous quarries, leading to shortage of soil–stone slag mixtures. The lime-treated soil can be easily obtained from the plain areas in Zhejiang Province, such as the waste soil from hydraulic reclamation in the Qiantang River, canal dredging, drilling in overpass construction, can be utilized. A comparison between the soil–stone slag mixture subgrade and the lime-treated soil subgrade in Jiaxing City is provided in Table 9.1.

H. Zhu et al., *Controlling Differential Settlement of Highway Soft Soil Subgrade*, SpringerBriefs in Applied Sciences and Technology, https://doi.org/10.1007/978-981-13-0722-5_9

Table 9.1 Comparison between soil–stone slag mixture subgrade and lime-treated soil subgrade in Jiaxing City

Comparison items	Soil–stone slag mixture	Lime-treated soil subgrade
Material source	Cutting into mountains is necessary; it is mainly transported from other places. An increasing shutdown of quarries, which exacerbates purchasing. There is impact from rain and typhoon, making it difficult for ships to enter and exit sites	Soil is obtained at the local level, particularly, under the background of urban reconstruction. Channels are also provided for receiving the subsequent waterway earthwork
Environmental impact	Mountains and environment are damaged	Favorable for disposing urban development earthwork, reducing the impact of stacking on land; conducive to treatment of sludge from rivers
Degree of compaction	Unstable, non-uniform	Stable, uniform
Strength	High dissociation, non-uniform	Low dissociation, uniform
Water stability	Subgrade clearance is large; subgrade is pervious to water; water is present in the subgrade for a long time during operation	Subgrade clearance is small; in a closed state and impervious to water
Subsequent operation	Unstable, large settlement deformation	Stable, small settlement deformation
Cost	Affected by seasons and the shutdown of quarries; the prices of raw materials highly fluctuate; price increase represents the general trend	Relatively stable; cost is controllable and is not higher than that of soil–stone slag mixture
Climatic impact	Construction is less affected by climate	Construction is greatly affected by rainy seasons, progress is subject to a certain impact

9.2 Applications of the Lime-Treated Soil Subgrade in Jiaxing and Tongxiang

(1) Inspirations from the cases that involve the application of lime-treated soil in bridgehead subgrades in Jiaxing City

Adjacent to Majiabang in Jiaxing City, the Nanjiao River Bridge on National Highway 320 (G320), which is a first-class highway, was designed by Suzhou Municipal Works Design Institute and built by Jiaxing Waterway Construction Co., Ltd. as the owner.

This project involves ten connecting lines (ramps). The lime-treated soil subgrade is approximately 120,000 m^3. According to the design requirements for different sections, construction was performed using 4, 6, and 8% lime-treated soil;

the 4% lime-treated soil was installed between the bottom of the highway trough at the filling section and the cushion layer and in the sidewalk subgrade and base layer. The maximum compaction thickness of each layer does not exceed 20 cm. Six percent lime-treated soil was installed in the cushion layers constructed after 30 cm of excavation at the top of the soft subgrade's mixed piles, whose construction was performed by two layers. Eight percent lime-treated soil was installed 80 cm from the road trough at the filling section, whose construction was conducted by five layers; the thickness of each compacted layer was 16 cm. To avoid performing construction in unfavorable seasons and rainy seasons, subgrade construction was conducted from May, 2006 to May, 2007. The minimum air temperature was 5 °C and the construction ceased one month prior to freezing to prevent the surface layer of the lime-treated soil from being frozen.

Seven years after this lime-treated soil section was opened to traffic, the application effect was apparently better than the same remaining highway using the soil–stone slag mixture subgrade. This finding provides excellent inspirations to us.

(2) Pilot projects of lime-treated soil subgrades in Tongxiang

To systematically study the effect of lime-stabilized soil on building the subgrade of highway, construction is performed on the following two fronts: first, various construction processes, including the mix proportion in the lime-treated soil subgrade, construction machinery, control and quality test, are determined according to the soil properties and characteristics in Jiaxing area. This study has yielded staged results; since the second half of 2013, the lime-treated soil subgrade has been successfully installed on Tongjiu Highway in Tongxiang City, Provincial Highway 203 (S203) in Haining City, which has produced significant effect and effectively improved the subgrade stability. Second, given the steel slag from Jiaxing Steel Works (steel slag is a light material between lime and fly ash), the composite soil modified by steel slag and lime is investigated, and comprehensive utilization can solve the occupation of massive farmland and land by existing steel slag and the resulting environmental pollution. Currently, the Jiaxing Highway Department has cooperated with Southeast University to analyze and test steel slag; the indoor test has been proven, and it will be applied in the test sections in Tongxiang during the next stage.

On December 13, 2013, Zhejiang Highway Administration held a technical exchange meeting on lime-stabilized soil subgrade in Tongxiang. This meeting produced a significant effect and pinpointed the direction for further application of lime-stabilized soil subgrade. Jiaxing Highway Department will continue to conduct a relevant study and constantly make this technology more applicable and scientific to lay an excellent foundation for large-scale application of this technology.

9.3 Application of Lime-Treated Soil Subgrade in Provincial Highway 203 in Haining City

(1) Project overview

The length of Provincial Highway 203 is 33 km long within Haining City. Phase I (Huyan Highway—Provincial Highway 01) spans 11 km and was completed in 2010; Phase II (Maqiao, Haining—Jianshan) is a 19.7 km reconstruction project, whose construction began in September, 2013. This provincial highway is a first-class highway with four bidirectional lanes; the width of the subgrade is 24.5 m, and the bridge is designed with six lanes.

(2) Background and significance

    (1) Purchasing the raw material soil–stone slag mixture is difficult. For the Phase II project of Provincial Highway 203, whose construction officially began in September, 2013, soil–stone slag mixture has been adopted for the entire subgrade according to the original design. Currently, an increasing number of quarries have been shut down. Although mountains are available in Jianshan, quarrying cannot be conducted. All existing quarries in Haining have stopped production. Purchasing the raw material is extremely difficult. Construction has been suspended to wait for the raw material.

    (2) Soil exists on the site but cannot be employed. The area within the scope of approximately 7 km at the right side of the south section of Provincial Highway 203 is used for hydraulic reclamation from the Qiantang River. Obtaining soil is convenient but the soil is silty soil in a liquid-plastic state. According to *Technical Specifications for the Construction of Highway Subgrades* (JTG F10–2006) ("the Specification"), **silty soil is unsuitable for being directly used in building a roadbed and cannot be directly used to build the subgrades that are soaked in water and the roadbeds in frozen areas.** Can this silty soil be converted into the subgrade filler through technical treatment? If successful, this treatment will provide inexhaustible highway material for use in the area along the Qiantang River, which saves energy and reduces emissions, as well as guarantees progress in this project.

    (3) Inspirations from a survey on Zhangjiagang in Jiangsu Province. The soil quality in Zhangjiagang is similar to that in the Qiantang River. The difference is that the soil in Zhangjiagang has a slightly higher clayey level. In Jiangsu, the raw material is obtained at the local level with 5–8% lime added into the undisturbed soil to stabilize the soil and 12% lime added as the subbase layer of pavement to replace the subbase layer made of cement stabilized macadam; its quality is good (Figs. 9.1 and 9.2). Based on a complete solicitation of opinions from various sources, three highway test sections were finally chosen, with different test schemes adopted to determine the actual effect to satisfy the performance requirements for subgrade material.

Fig. 9.1 Current situation of subgrade base in Zhangjiagang

Fig. 9.2 Lime-treated subgrade

Table 9.2 Physical mechanical indicator one concerning the soil samples from the test highway sections

Highway section	Geotechnical property	Natural moisture content %	Natural wet density g/cm^3	Natural void ratio	Liquid limit	Plastic limit
I	Silty clay	28.4	1.83	0.998	37.2	24.4
II	Silty clay	29.3	1.85	0.982	36.8	22.3
III	Sandy silt	37.9	1.89	0.82	26.7	19.5
	Clayey silt	39.2	1.87	0.865	28.7	20.6

Table 9.3 Physical mechanical indicator two concerning the soil samples from the test highway sections

Highway section	Geotechnical property	Plasticity index	Liquidity index	Cohesion (kPa)	Internal friction angle (°)
I	Silty clay	15.2	0.76	27	8.8
II	Silty clay	16.5	0.83	37	15.5
III	Sandy silt	6.2	1.17	6	27.1
	Clayey silt	8.1	1.06	8	28.6

(3) Test schemes

The soil for the subgrade in the Provincial Highway 203 reconstruction project primarily consists of clay and silt (generated by hydraulic reclamation from the Qiantang River). Clay is characterized by its low moisture content, high plasticity index and difficult compaction, whereas silt is characterized as with high moisture content, low plasticity index and is easy to scour and unstable. If clay and silt are not treated, their CBR, degree of compaction and stability will fail to comply with the specifications; thus, they were hardly used in the previous highway subgrades in Zhejiang.

Clay and silt contain SiO_2 and SO_4, etc. With the addition of lime and cement, their effective constituents CaO and $Ca(OH)_2$ engaged in a chemical reaction with CO_2 in the air to generate calcium silicate gel, $CaCO_3$, $CaSiO_2$, and $CaSO_4$. As a result, first, increase the degree of density; second, reduce gaseous and liquid porosity; third, cut off pores; and fourth, enhance the overall strength.

Based on the lime mixing tests (5, 6, 8, and 10%) on three trail highway sections, the soil quality was improved. An on-site test was conducted with respect to the degree of compaction, CBR and deflection to examine whether the improved performance complies with the indicator requirements in the specifications.

(1) On-site physical indicator test of soil samples

The physical mechanical indicators concerning the soil samples from test highway sections are listed in Tables 9.2 and 9.3.

Table 9.4 Requirements for the design of indicator the degree of compaction at the clay highway sections

Distance from roadbed top surface (cm)	Degree of compaction (%)	Remarks
0–80	≥96	0–30 cm Lime 8%, transverse slope 2%
		30–60 cm Lime 6%, transverse slope 3%
		60–80 cm Lime 6%, transverse slope 3%
80–95	≥94	80–95 cm Lime 5%, transverse slope 3%
95–110	≥90	95–110 cm Lime 5%, transverse slope 3%
>110	≥87	Lime 5%, transverse slope 3%

Table 9.5 Requirements for the design indicator of the degree of compaction at the silty soil highway sections

Distance from roadbed top surface (cm)	Degree of compaction (%)	Remarks
0–80	≥96	0–30 cm Lime 10%, transverse slope 2%
		30–60 cm Lime 8%, transverse slope 3%
		60–80 cm Lime 8%, transverse slope 3%
80–95	≥94	80–95 cm Lime 6%, transverse slope 3%
95–120	≥90	95–110 cm Lime 6%, transverse slope 3%
>120	≥87	Lime 6%, transverse slope 3%

Table 9.6 CBR and subgrade top surface deflection requirements

Filler application site (cm)	Minimum strength of filler CBR (%)	Top surface deflection (1/100 mm)
0–30	8	≤266
30–80	5	
>80	4	≤331

(2) Requirements for the design indicator of the degree of compaction

The requirements for the design indicator of the degree of compaction at the test highway sections are shown in Tables 9.4 and 9.5.

(3) CBR and subgrade top surface deflection requirements

The requirements for CBR and subgrade top surface deflection at the test highway sections are listed in Table 9.6.

(4) Lime quality requirements

Grade III and above of quick lime satisfying the specified technique specification should be employed; its effective CaO and MgO content should exceed 70%, and its fineness should exceed 80%.

Table 9.7 Summary of indoor test data concerning the best moisture content and degree of compaction at the clay highway sections

Lime content (%)	Best moisture content (%)	Maximum dry density (g/cm^3)	Remarks
5	14.90	1.644	
6	14.70	1.637	

Table 9.8 Summary of indoor test data concerning the best moisture content and degree of compaction at the silt highway sections

Lime content (%)	Best moisture content (%)	Maximum dry density (g/cm^3)	Remarks
5	13.00	1.476	Highway through 1 m soil
6	12.50	1.538	Highway through 0.8 m soil
8	15.40	1.579	Highway trough 0.5 m soil
10	14.80	1.631	Highway trough 0.3 m soil

Table 9.9 Indoor CBR test on the test highway sections after lime mixing

Highway section	Soil quality	5% lime	6% lime	8% lime
I	Silty clay	10.0	12.6	
II	Silty clay	10.0	12.6	
III	Sandy silt	5.0	10.5	12.1

Table 9.10 Summary of test on the degree of compaction at the test highway sections after lime mixing

Highway section	Soil quality	5% lime		6% lime		8% lime	
		Number of points	Degree of compaction	Number of points	Degree of compaction	Number of points	Degree of compaction
I	Silty clay	16	91.2	8	97.3		
II	Silty clay	24	90.7	8	97.2		
III	Sandy silt			53	90.7	62	96.7

(5) Test results

The test results concerning the test highway sections are shown in Tables 9.7, 9.8, 9.9, 9.10, 9.11, 9.12 and 9.13.

Based on test and analysis of data (CBR, degree of compaction, and stability), it is believed that lime mixing can produce a suitable effect on the improvement of soil quality. In particular, the clay quality significantly improved, and significant improvement in the base compaction degree, overall strength and water stability was achieved, as shown in Figs. 9.3, 9.4, 9.5 and 9.6. Therefore, after the undisturbed soil is added with a certain proportion of quick lime, it can serve as a subgrade filler.

Table 9.11 Summary of test on deflection at the test highway sections after lime mixing

Highway section	Soil quality	Number of layers	Number of points	Average value	Representative value
I	Silty clay				
II	Silty clay				
III	Sandy silt	3	32	223.78	270.02

Table 9.12 Summary of the (indoor) test on unconfined strength of the slit subgrade

Lime content	7d	14d	21d	28d
6% lime	0.07	0.26	0.30	0.32
8% lime	0.13	0.28	0.36	0.41
3% lime + 3% cement	0.41	0.67	0.8	0.86
3% lime + 5% cement	0.65	0.85	1.14	1.22

Table 9.13 Silt CBR test data

Lime ratio	CBR
3% cement + 3% lime	28
4% cement + 4% lime	34
5% cement + 5% lime	42

(4) Comprehensive analysis

(1) As indicated by the test on the subgrades after the undisturbed soil was treated by lime at the clay and silt highway sections, the minimum strength CBR of the lime-treated soil filler and the subgrades' degree of compaction can comply with specifications. They were apparently better than the soil—stone slag mixture subgrades, and the total strength increased with time.

(2) The lime mixing techniques are adopted according to the soil quality in Haining to ensure a feasible highway performance. This finding has significant realistic guidance for solving the current severe shortage of soil–stone slag mixture and slow progress in this project.

(3) With regarding to the unstable characteristics of silty soil subgrade, such as easy to scour and collapse, the following treatment measures are performed to achieve excellent effects:

(a) Strength drainage. Excavate deep side ditches in a timely manner on both sides when building the silty soil subgrade. Facilitate discharge of surface water and underground water, and prevent the subgrade from rain erosion to ensure that the moisture content in the subgrade dose not exceed the best moisture content in soil.

Fig. 9.3 Current situation of soil subgrade after base excavation at the silt sections

(b) Mitigate the side slope. Moderately widen the subgrade; mitigating the side slope produces soil in a natural state that uneasily collapses; increase the compaction width at each side of the subgrade by 50 cm to reserve the scour width; and maintain and safeguard the stability of the subgrade.

(c) Properly increase the transverse slope. 3–4% crown is conducive to timely discharging the rainfall within the subgrade to the outside of the subgrade to avoid local downspouting and prevent water from infiltrating the subgrade.

Fig. 9.4 Effect of compaction after lime mixing at the silt sections

    (d) Protect the side slope in time. After improvement of the section along the river, immediately take measures such as a mortar protective slope and protective wall to prevent rainfall from directly eroding the soil subgrade.

    (e) Adopt a comprehensive lime-cement stabilization method to conduct indoor and outdoor tests, and significantly improve the consolidation effect, strength and water stability of the lime-treated soil subgrade.

(4) Addressing the issue of soil source is critical. The design should consider the subgrade height, the soil obtaining sites in the surrounding areas should be surveyed to properly increase the area of land acquisition and coordination with the Municipal Slag should be ensured to designate the discarded soil sites.

Fig. 9.5 Changes in soil quality after addition of lime in clay soil, slaking and the first tedding

According to the test on the subgrades after the undisturbed soil was treated by lime, the subgrade's CBR, degree of compaction and settlement comply with the specification. They are apparently superior to the soil–stone slag mixture subgrades.

Based on the success of the test highway sections, the lime-treated improved soil was backfilled at the 8 km section of Provincial Highway 203. This project was completed and opened to traffic in January, 2015. The test results are shown in Table 9.14; the subgrade is excellent in terms of all indicators. Currently, it has been under trial operation for near two years. The entire subgrade is smooth and stable without bumps at the bridgeheads. Moreover, about 350,000 m^3 soil–stone slag mixture has been saved, which is equivalent to a 50 m-high hill that covers an area of about 0.67 hectares.

Table 9.14 Test results concerning the top surface deflection at the subgrade of Provincial Highway 203

Pile No.	Design value	Average value	Representative value	Note
K0+000–K1+000 left	266	97	228	Slag section
K0+000–K1+000 right		133	260	Slag section
K1+000–K2+000 left		116	257	Slag section
K1+000–K2+000 right		111	246	Slag section
K2+000–K3+000 left		119	251	Slag section
K2+000–K3+000 right		114	252	Slag section
K3+000–K4+000 left		129	249	Slag section
K3+000–K4+000 right		121	234	Slag section
K4+000–K4+500 left		84	170	Lime section
K4+000–K4+500 right		83	160	Lime section
K4+500–K4+600 left		100	177	Lime section
K4+500–K4+600 right	266	98	173	Lime section
K4+600–K4+719.48 left		144	216	Slag section
K4+600–K4+719.48 right		140	210	Slag section
K4+764.52–K5+000 left		97	160	Lime section
K4+764.52–K5+000 right		99	155	Lime section
K5+000–K6+000 left		92	180	Lime section
K5+000–K6+000 right		97	193	Lime section
K6+000–K7+000 left		40	100	Lime section
K6+000–K7+000 right		37	90	Lime section
K7+000–K8+000 left		42	127	Lime section
K7+000–K8+000 right		44	131	Lime section
K8+000–K9+000 left	266	106	201	Lime section
K8+000–K9+000 right		108	217	Lime section
K9+000–K10+000 left		74	152	Lime section
K9+000–K10+000 right		76	153	Lime section
K10+000–K10+623.42 left		81	157	Lime section
K10+000–K10+623.42 right		80	177	Lime soil section
K10+696.52–K11+000 left		98	186	Slag section
K10+696.52–K11+000 right		90	156	Slag section
K11+000–K12+000 left		140	200	Slag subgrade
K11+000–K12+000 right		144	201	Slag subgrade

Fig. 9.6 Overall effect after blending, spreading, and rolling compaction at the clay highway sections

(5) Problems and suggestions

    (1) The following problems exist in the subgrade built by the soil from the Qianjiang River (sludge becomes dry):

        (a) The soil within 5 cm on the surface layer is relatively loose and it is easy to cause "peeling" (Fig. 9.7).

        (b) Scouring occurs (Fig. 9.8).

        (c) Eliminating wheel marks is difficult (Fig. 9.9).

With regarding to the above problems, we are currently drawing the construction experience from other areas, and conducting indoor tests on comprehensive cement-lime improvement to produce the best effect. The test results are listed in Table 9.15.

Fig. 9.7 Difficulties in rolling compaction of "peeling"

Fig. 9.8 Scouring

As indicated by the tests (Table 9.15), after a certain proportion of cement was added to the silt from the bottom of the Qianjiang River, it significantly improved the consolidation effect, strength and water stability of the subgrade.

(2) When earthwork is used to build the subgrade, the area of land acquisition should be increased to obtain soil; subsequently, we can coordinate the entire environment by drainage through the side slope and greening improvement.

Fig. 9.9 Wheel marks

Table 9.15 Summary of the (indoor) test on the unconfined strength of the silt subgrade (unit: MPa)

Lime ration	7d	14d	21d	28d
6% lime	0.07	0.26	0.30	0.32
8% lime	0.13	0.28	0.36	0.41
5% lime + 1% cement	0.20	0.28	0.31	0.34
4% lime + 2% cement	0.35	0.38	0.41	0.48
3% lime + 3% cement	0.41	0.67	0.8	0.86
3% lime + 5% cement	0.65	0.85	1.14	1.22

(3) The undisturbed soil is treated by lime for improvement to achieve highway material performance that is consistent with the requirements and thus solve the current severe shortage of soil–stone slag mixture for the subgrade. This treatment has significant realistic guidance for smooth progress in this project. Although the lime-treated improved soil is greatly affected by weather, it can be obtained at the local level to save resources. Therefore, the lime-treated improved soil is obviously superior to soil–stone slag mixture, which deserves to be summarized and widely applied. Further in-depth study of the improved material should be conducted.

Chapter 10
Practice in Treatment of the Bumps at Bridgeheads on the Existing Highway in Taizhou City

10.1 Treatment Measure of Combining Light Materials with the Lower Diaphragm Plates for the Soft Soil Subgrade

The foam bead concrete consists of foam beads (EPS particles), medium-coarse sand, macadam, cement, and water at a certain designed mix proportion through blending. All materials used in the tests are chosen and obtained nearby. The foam beads (EPS particles) are polystyrene spherical particles with a diameter of 3–5 mm and an apparent density of 5–20 kg/m^3. The cement is ordinary Portland cement PO32.5. For the aggregate, the sand fineness modulus in the fine aggregate ranges from 3.0 to 2.3. The size distribution of macadam particles in the coarse aggregate ranges from 5 to 21.5 mm. The admixtures consist of common early-strength water-reducing agents, microsilica and fly ash.

The foam concrete is a light material that is made by full mixing and stirring of hardener (cement), water, foam and other admixtures at a certain mix proportion. The concrete is characterized by light weight; the adjustability of weight per unit volume and strength; high fluidity; post-consolidation stability; excellent constructability; high durability and superior environmental protection properties.

The finished polyethylene foam block (EPS block) replaces the soil behind the abutments to reduce the soil pressure on the abutments.

The treatment of combining light materials with the lower diaphragm plates for the soft soil subgrade is shown in Fig. 10.1.

© Springer Nature Singapore Pte Ltd. and Zhejiang University Press 2019
H. Zhu et al., *Controlling Differential Settlement of Highway Soft Soil Subgrade*,
SpringerBriefs in Applied Sciences and Technology,
https://doi.org/10.1007/978-981-13-0722-5_10

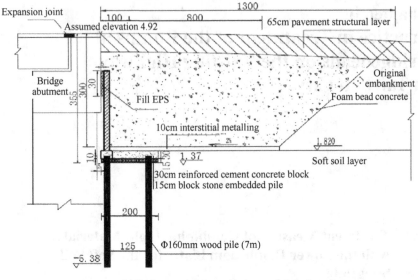

(a) Foam concrete + lower diaphragm plate (unit: cm)

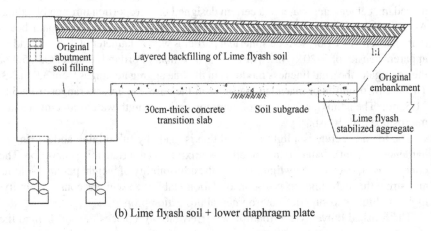

(b) Lime flyash soil + lower diaphragm plate

Fig. 10.1 Treatment measure of combining light materials with the lower diaphragm plates for the soft soil subgrade

10.2 Soft Soil Subgrade Grouting Measure

(1) Deep grouting

The soil behind the abutment is drilled at a depth of 6–8 m. The drilling plane features quincunx; grouting is conducted by A chemical quick setting slurry for consolidation of soil subgrade, as shown in Figs. 10.2 and 10.3.

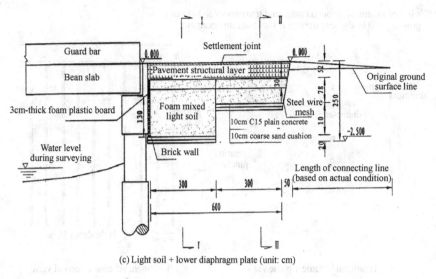

(c) Light soil + lower diaphragm plate (unit: cm)

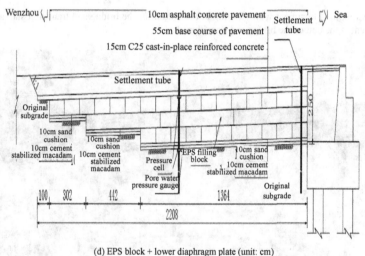

(d) EPS block + lower diaphragm plate (unit: cm)

Fig. 10.1 (continued)

(2) Shallow grouting

With a drilling depth of approximately 60 cm, water glass, aluminum powder and other quick setting slurry and micro-expansive slurry are used to grout the shallow subgrade. They can reinforce the subgrade. In particular, they can produce an excellent effect on the soft soil subgrade that is built by the soil–stone slag mixture, as shown in Figs. 10.4 and 10.5.

During 2007–2014, the soft soil subgrade grouting technique was successfully applied to numerous highways and urban roads, including Mapu New Bridge,

Combined treatment through bridge end transition slab bottom
grouting and bridge end transition slab end reinforcement

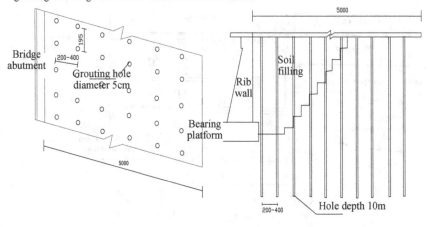

Treatment scheme 1 plane view   Treatment scheme 1 vertical view

Fig. 10.2 Combined treatment with grouting at the bottom of the bridge end transition slab and reinforcement at the end of the bridge end transition slab

Fig. 10.3 Deep grouting in the soft soil subgrade

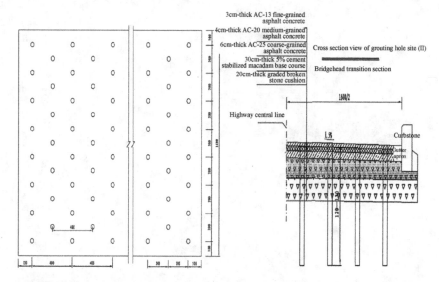

Fig. 10.4 Arrangement of subgrade compaction grouting holes (unit: cm)

Fig. 10.5 Shallow grouting in the soft soil subgrade

Kaifaqu Bridge, Xinsihao Bridge, Ximen Overpass, Yongning River Bridge, Feng-shan Bridge, the collar beam at National Highway 104 that overpasses Ningbo-Taizhou-Wenzhou Lingjiang River Bridge, multiple bridges at the extension of Provincial Highway 82, the subgrade of Luqiao Dongfang Avenue, and Jiaojiang Kaifa Avenue.

Chapter 11
Practice in Treatment of the Bumps at Bridgeheads on the Highways in Wenzhou City

11.1 Deep Grouting for Treating the Bumps at Bridgeheads at Ouhai Section of National Highway 104

With a total length of 10 km, the Ouhai section of National Highway 104 in Wenzhou is a typical soft soil section. After this section was opened to traffic after reconstruction in 2000, severe differential settlement developed in the soft soil subgrade, causing very severe bumps at the bridgeheads. The average daily mixed traffic volume at this section ranges from 50,000 to 60,000 vehicles. This section has 13 bridges. The bridgeheads need to be paved each year. To effectively solve the bumps at bridgeheads, in 2008, Xiacheng No. 2 Bridge was chosen as the test project, in which four different methods were adopted to treat the corresponding four bridgeheads.

Xiacheng No. 2 Bridge was built in 2001. The bridge end transition slabs were placed behind the bridge abutment. The width of the subgrade of the bridgehead that connects the line is 25.5 m. The width of the middle green belt is 2.0; the width of the lanes is 2 m × 7.75 m; the width of auxiliary lanes is 2 m × 3.5 m. After this bridge was completed, natural settlement existed in the soil at the bridgehead. After 6 years later in 2007, the total settlement was about 85 cm, and asphalt was added behind the bridge abutment for five times, which was about 10–15 cm for each time. The bridge end transition slabs have shifted to leave voids. Although a significant amount of manpower and materials was put into maintenance, the bumps at bridgehead were mitigated to some extent in the early period of maintenance. However, as time passes, the bridge end transition slabs shifted to result in voids, the bumps at bridgehead became increasingly severe.

With this bridge as the background, we surveyed the bridges with the bumps at bridgeheads, analyzed the disease form and disease mechanism, and proceeded from the perspective of highway maintenance to avoid extensive excavation, shorten the maintenance period and mitigate the impact on highway traffic. We proposed the well-targeted rectification measures and implementation plans to control the bumps at bridgeheads within the permissible range.

© Springer Nature Singapore Pte Ltd. and Zhejiang University Press 2019
H. Zhu et al., *Controlling Differential Settlement of Highway Soft Soil Subgrade*,
SpringerBriefs in Applied Sciences and Technology,
https://doi.org/10.1007/978-981-13-0722-5_11

Four treatment methods are described as follows:

(1) The rapid repair technique for soil backfilling behind abutments through deep grouting reinforcement (DGR)

The grouting length was 50 m; the grouting hole size was 5 cm; the grouting holes were horizontally arranged in five rows; the spacing of vertical hole ranged from 2 to 4 m; the grouting depth was 10 m (the grouting depth depends on the requirement of the subgrade's bearing capacity and the degree of soil strength), the grouting pressure ranged from 0.5 to 1 MPa. The bridge was opened to traffic 12 h after grouting was completed.

(2) Combined treatment including removal of bridgehead pedal and grouting in the surface of soil backfilling behind abutment

The construction process is the same as in (1).

(3) Grouting at the bottom of the bridge end transition slab and vertical reinforcement at the end of the baffle plate

Grouting reinforcement was conducted to be the same as soil backfilling behind abutment. The grouting range was 1.5 m larger than the length of the bridge end transition slab. The grouting hole size was 5 cm; the grouting holes featured a triangular arrangement. The spacing was 150 cm × 150 cm. Shallow grouting was carried out at the bottom of the bridge end transition slab, whereas vertical grouting was conducted at the end of the bridge end transition slab. The grouting depth was 10 m; the grouting pressure ranged from 0.5 to 1 MPa. The bridge was opened to traffic 12 h after grouting completion.

(4) The bridgehead connecting slope was adjusted by the cement stabilized layer; and the surface layer was paved with bituminous concrete. This method was primarily employed as a comparative scheme.

We observed settlement at the connecting line section of the connecting slope at the head of the test bridge, and arranged three horizontal observation points 1, 5 and 9 m away from the bridge abutment. After one year, the total settlement was approximately 1 cm, which suggests a significant effect from the deep grouting reinforcement method.

In response to the bumps at the bridgeheads of the remaining 12 bridges on National Highway 104 (Table 11.1), the deep grouting reinforcement technique was adopted in 2009 to deal with the soft subgrades at the bridgeheads.

After subsequent four years, in June, 2013, the settlement at the bridgeheads that was reinforced through grouting was measured, which indicated that the settlement was basically stable with the most settlement of the bridgeheads ranged from 13 to 15 mm, and the overall effect was good.

Table 11.1 Detailed table of 12 grouted bridges at Ouhai section of G104 Highway

Highway name	Bridge name	Bridge site pile No.	Bridge length (m)	Remarks
G104	Yuhu	K1917+168	117	
G104	Ehu	K1918+392	52	Left and right sides
G104	Xiadun	K1919+184	84	Left and right sides
G104	Wangzhai	K1920+172	85	Left and right sides
G104	Xiazhang	K1920+738	22.6	Left and right sides
G104	Jiangzhai	K1921+516	70.6	Left and right sides
G104	Xiacheng I	K1922+103	84	Left and right sides
G104	Linshan I	K1923+232	68	Left and right sides
G104	Linshan II	K1923+726	61.7	Left and right sides
G104	Zhuxi II	K1925+246	36	Left and right sides
G104	Zhuxi I	K1924+812	62	Left and right sides
G104	Shen'ao	K1925+973	30	Left and right sides

11.2 Measure for Treating the Soft Subgrade Transition Position of the Elevated Bearing Platform in a Highway Project in Oujiangkou, Wenzhou City

A highway project in Oujiangkou, Wenzhou City, Zhejiang Province is located in the marine depositional plain, where the terrain is flat and wide. The entire line is a soft soil one. The surface is a hard shell layer. The engineering properties are relatively poor. The subgrade's bearing capacity is $fa_0 = 80 - 90$ kPa. The upper thick layer is a mucky soil layer, the maximum thickness is 40 m, the engineering properties are extremely poor, and the subgrade's bearing capacity is $fa_0 = 40 - 80$ kPa, the moisture content ranges from 49.7 to 69.8%, the void ratio ranges from 1.502 to 1.937, the compression coefficient ranges from 0.790 to 2.300 MPa, and the compression modulus ranges from 1.47 to 3.19 MPa. The middle and lower parts are clay layers in a soft plastic—plastic state, the engineering properties are relatively poor—general, the subgrade's bearing capacity is $fa_0 = 110 - 180$ kPa.

A portal framed pier is placed on the side strip of this project because a track is added at the access section of the S2 line in Wenzhou intracity railway project. Based on consultation with the S2 Line Design Institute, the width pf the portal framed pier is 2 m. Given a certain safe distance, the width of the side strip is 3 m. Typically, the cross section layout is expressed as follows: 0.5 m soil shoulder + 2.75 m auxiliary road + 3 m side strip + 0.5 m marginal strip + 3 × 3.75 m lane + 0.5 m marginal strip + 6 m medial strip + 0.5 m marginal strip + 3 × 3.75 m lane + 0.5 m marginal strip + 3 m side strip + 2.75 m auxiliary road + 0.5 m soil shoulder = 43 m.

The S2 line bearing structure is located below the lanes of this project. With the impact from the pile foundation, the settlement in the pavement within a certain scope of the pavement especially near the pile foundation will be relatively small; cracks easily form within the general section pavement structure. Therefore, consideration is made to improve the compatibility of the settlement deformation in the subgrade at both sides of the pile foundation in the S2 section of this project, reducing the settlement of the subgrade to prevent the negative friction of settlement from adversely affecting the pile foundation. Given the positional relation between the S2 line and this project, a construction treatment analysis is conducted on the following two conditions.

(1) The section shared by this project and intracity railway S2 line, as shown in Fig. 11.1.

Only part of the pavement at the medial strip at the shared section is affected by the S2 pier; in principle, pile foundation + reinforced concrete sash are adopted. Within the area, the pile foundation of the soft soil subgrade is provided with cement-mixed piles or prestressed pipe piles according to the thickness of the soft soil layer. Generally, when the thickness of the soft soil layer is 15–20 m, cement-mixed piles are adopted. When the thickness of the soft soil layer is relatively large, prestressed pipe piles are adopted. Cement-mixed piles are installed near the pier; and the impact of the construction disturbance is relatively small. In practical construction, the minimum distance between the edge of the mixed piles or the edge of the pipe pile cap and the edge of the bearing platform is 50 cm. For the entry of the S2 pier pile foundation or the bearing platform into the scope of the lanes in this project, attention should be paid to avoidance during pile driving, the pile location should fall outside the scope of the bearing platform, and the minimum distance between the edge of pile cap and the edge of the bearing platform is 50 cm.

During the settlement of the soft soil subgrade, a large settlement deformation difference easily causes cracks in the pavement structure. This scheme is performed with pile foundation treatment, and the overall deformation compatibility of the subgrade is controlled. After the construction of the pile foundation for the soft subgrade is completed, the reinforced concrete sash is placed on the top of the pile foundation. The sash size is determined according to the pile spacing; the design thickness generally ranges from 20 to 24 cm; and the specific location is treated by leveling the cement mortar according to the elevation of the pile foundation top.

Given that a certain amount of soft soil settlement will exist in the middle of the sash after treatment, one steel-plastic composite geogrid is added between the roadbed bottom and the geogrid top surface; one glass fiber grid is added at the bottom of the asphalt surface courses on the pavement; one side of grid is extended to the edge of the medial strip; and the reaming sides are extended beyond the bearing

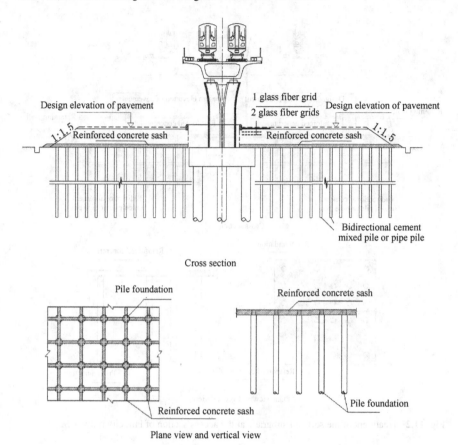

Cross section

Plane view and vertical view

Fig. 11.1 Treatment of the soft soil subgrade at the shared section

platform for 2 m. With respect to the steel-plastic composite geogrid, its maximum elongation at break is required to be 3%; and its minimum horizontal and vertical tensile strength is 100 kN/m. With regard to the glass fiber grid, its bidirectional tensile strength is required to be not less than 50 kN/m, and its maximum elongation at break is required o be 3%.

(2) The access section of intracity railway S2, as shown in Fig. 11.2.

The section is subject to the boundary line of land; thus, the S2 pier pile foundation bearing platform cannot be completely located within a medial strip or side strip. The bearing platform will be extended to the scope of lanes in this project. The lane subgrade is integrally located between the S2 pile foundations. With the influence of the bearing platforms of the three rows pile foundation, the scope of impact is more extensive. Therefore, the construction effect should be more strictly controlled. To reduce the subgrade settlement difference inside and outside the bearing platforms,

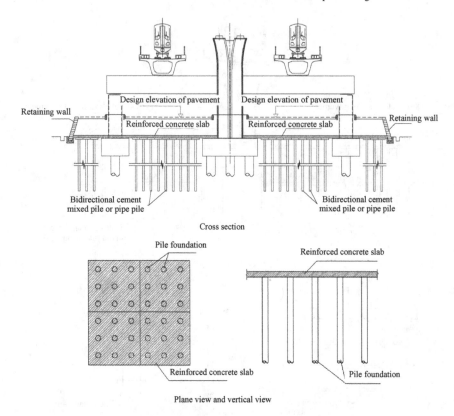

Fig. 11.2 Treatment of the soft soil subgrade at the access section of intracity railway S2

pile foundation + reinforced concrete slab are adopted. The pile foundation mainly adopts cement-mixed piles.

During practical construction, the minimum distance between the edge of the mixed pile and the edge of the bearing platform is 50 cm. Since the S2 pier pile foundation or the bearing platform will enter into the scope of lanes in this project, special attention should be paid to evasive action during pile driving, and the pile location should fall outside the scope of the bearing platform. In this scheme, the reinforced concrete slabs are placed on the top of the pile foundation; the design thickness of concrete is 20–24 cm. Slab bodies undergo rectangular slab-by-slab treatment according to the pile spacing, the area of a single slab should be within the range of 24–30 m². The role of the reinforced concrete slab is greater than that of the reinforced concrete sash in the previous scheme. According to the preliminary results, both roles mitigate the highway settlement deformation by improving the soft soil subgrade deformation compatibility under the action of pile foundation. The difference is that the reinforced concrete slab structure evenly distributes the upper load in the soil between pile foundation and the pile. Therefore, the pile top does not produce the soil arch effect, and the embankment soil filling height can also be properly optimized and reduced.

No connecting pieces are placed between the divided reinforced concrete slabs. However, a steel-plastic composite geogrid should be placed at the dividing joint to reduce the crack reflection between the slab widths, which is caused by differential settlement. In addition, a geogrid should be placed in the position of the S2 pier; when it is placed, attention should be paid to reserving the site for the portal framed pier bearing platform.

Chapter 12
Practice in Treatment of the Bumps at Bridgeheads of the Highways in Ningbo City

12.1 Practice in the Lime-Treated Soil Subgrades of the Highways in Ninghai Urban Areas

The highway alteration project of the section from Meilin to Shanheling in Ninghai urban areas involves a line with a total length of 11.235 km. The section is designed according to the standard for the first-class highway with 4–6 lanes; the design speed is 80 km/h; and the subgrade width is 26.5–45 m. The transportation of earthwork is restricted during highway excavation in urban areas. In addition, the outsourced fillers are also prohibited due to the difficulty in transportation. Therefore, it is more appropriate to use the undisturbed soil to carry out the construction after the undisturbed soil is treated by lime. Specific judgment can be made from an economical and environmental protection perspective, as shown in Tables 12.1 and 12.2.

To guarantee the filling quality of the lime-treated soil subgrade, a lime-treated soil subgrade test was conducted from K7+980 to ZK8+150 section to determine the reasonable technical parameters of construction.

The soil in this subgrade includes the filled soil, farming and planting soil, silty clay soil, and macadam-containing silty clay soil. The lime-treated soil construction is primarily employed in the subgrade filling at the section with low filling and shallow excavation. The test section is divided into the upper roadbed and the lower roadbed. The upper roadbed (pavement subbase 0–40 cm) is filled with 6% lime-treated soil in two layers, whereas the lower roadbed (pavement subbase 40–80 cm) is filled with 5% lime-treated soil in two layers.

(1) On-site collecting of soil samples

In the subgrade of the ZK7+980–ZK8+150 section, the maximum dry density is 1.78 g/cm^3, the best moisture content is 12.5%, the liquid limit is 35.7%, the plastic limit is 20.3%, and the plasticity index is 15.4. The length of the section is 170 m, the width of the construction is 23.5 m, the compaction thickness of each layer is 20 cm, and the lime grade is Grade III. The standard test on 5% lime-treated soil is

© Springer Nature Singapore Pte Ltd. and Zhejiang University Press 2019
H. Zhu et al., *Controlling Differential Settlement of Highway Soft Soil Subgrade*,
SpringerBriefs in Applied Sciences and Technology,
https://doi.org/10.1007/978-981-13-0722-5_12

Table 12.1 Comparison in economy between the lime-treated soil and the outsourced soil-stone slag mixture

Scheme	Price(yuan/m^3)	Remarks
Scheme one: outsourced soil-stone slag mixture	72.09	The prices of the outsourced soil-stone slag mixtures in similar projects: Xiangxi Line 70.7 yuan/m^3; Fengchaling Phase III 73.47 yuan/m^3. The average price is 72.09 yuan/m^3
Scheme two: lime-treated soil	$75.51 - 26.41 = 48.1$	The unit price of 5% lime-treated soil 75.51 yuan/m^3. No discarded soil is needed; Thus, the unit price of the discarded soil 26.41 yuan/m^3 is excluded. The actually increased unit price is 48.1 yuan/m^3
Conclusion	The lime-treated soil is relatively economical	

Table 12.2 Comparison in environmental protection between the lime-treated soil and the outsourced soil-stone slag mixture

Scheme	Advantages	Disadvantages
Scheme one: outsourced soil-stone slag mixture	Less construction machinery is involved; the construction process is simple; it is less affected by weather; the filling speed is relatively high	Blast mountain is necessary. Transport of the soil-stone slag mixture affects urban traffic and it also causes ecological destruction and air pollution
Scheme two: lime-treated soil	The discarded earthwork can be utilized; transport, discarded soil sites can be reduced. It does not cause ecological destruction, water and soil erosion and air pollution. It does not occupy land resources; it is relatively environment-friendly	More mixing machinery is involved; the construction process is relatively complicated. It is greatly affected by weather; the filling speed is relatively low
Conclusion	The lime-treated soil is relatively environment-friendly	

described as follows: the maximum dry density is 1.868 g/cm^3, and the best moisture content is 12.0%. The standard test on 6% lime-treated soil is expressed as follows: the maximum dry density is 1.872 g/cm^3, and the best moisture content is 11.8%.

(2) Construction process and quality control

   (1) Setting out
       Set out the king pile at the subgrade construction section, use a level gage to measure and calculate the construction width and elevation, use lime to

Fig. 12.1 On-site gray line frame

specify the center line and side lines, and carry out on-site pile driving, make points and draw lines to control the soil thickness.

(2) Raw material test

The natural loose dry density of slaked lime is 1230 kg/m³; the maximum dry density of 5% lime-treated soil is 1.868 g/cm³; and the best moisture content in lime-treated soil is 12.0%.

(3) Lime preparation

Deliver the packed slaked lime to the construction site.

(4) Soil delivery

Use excavator, loader, and soil delivery vehicle to transport the soil at the right side of the subgrade to the test section. Make the gray line frames on the site with a frame size of 5.88 m × 5 m. The specially-assigned personnel command and control vehicles to dump soil within the frames (Fig. 12.1).

(5) Spreading, tedding

Use an excavator to spread the soil and then preliminarily flatten the soil manually. The soil block particle size should be controlled at any time in this process. The soil blocks with a particle size more than 10 cm should be manually crushed. Cobble stones should be manually cleared, and the moisture content should be tested at any time for compliance with requirements. If the moisture content is high, tedding is required; if the moisture content is low, water spraying is required.

(6) Lime mixing

Testing personnel test the moisture content in the on-site soil samples. When the moisture content is higher than 2–3% of the best moisture content, start spreading the lime using a loader to load the lime, and transport the

Fig. 12.2 Lime mixing

lime to the test section (Fig. 12.2). After the lime is spread according to a certain quantity, evenly spread the lime onto the surface of the plain soil manually. Then use machines to mix the lime. After mixing, use a rotary cultivator to ted the mixed lime-treated soil until the lime-treated soil particles meet the specification requirements.

The results of an on-site test on lime-treated soil are described as follows:

(a) The lower roadbed 40–60 cm: 5% lime; the measured moisture content: 13.8, 14.6, 14.0, 13.2, 14.8, and 14.0%; the moisture content complies with the standard test requirements.

(b). The upper roadbed 20–40 cm: 6% lime; the measured moisture content: 14.7, 13.6, 14.2, 14.6, 14.6, and 13.5%; the moisture content complies with the standard test requirements (Fig. 12.2).

(7) Leveling

After lime-treated soil mixing, use a road roller to roll it at a time (because the fillers in the subgrade have more cobble stones, and leveler cannot be used). Use a bulldozer to manually conduct coarse leveling, and then use a vibratory roller to rapidly roll the initially leveled section for 1–2 times to expose the potential unevenness. Use an excavator to manually perform leveling. Prior to leveling, use a rotary cultivator to loosen the areas at a level that is 5 cm higher than the surface of the low-lying wheelmark areas. After leveling, use a vibratory roller to rapidly roll it. After rolling, restore the center line and side lines. The measuring personnel determine the thickness of the filling layer according to the corresponding coefficient of the loose paving material to drive side piles (spacing of 20 m). Specify the layer thickness on the side piles and use nylon ropes to connect them to control the filling width and thickness. No vehicle is allowed to pass through the site in the process of formation.

Fig. 12.3 Rolling compaction

(8) Rolling compaction

In the case of rolling compaction, start from the highway edge to the high-way center, and from the highway center to both sides. Conduct rolling compaction at different spacings back and forth, make wheel marks over-lap for 15–20 cm, and perform rolling compaction tests for 5, 6, 7, 8 times to obtain the best parameter combination. The first rolling and the last rolling are static rolling, whereas the other type of rolling is vibratory rolling (Fig. 12.3).

(3) Subgrade top surface rebound deflection test

The Benkelman beam is used to conduct a subgrade top surface rebound deflection test; the results are listed in Table 12.3.

(4) Comparison before and after subgrade formation

The comparison before and after subgrade formation is shown in Fig. 12.4.

12.2 Lower Diaphragm Plate Treatment Measure for the Soft Subgrade at the Bridgehead in the Highway Project in Zhenhai District

The project of a highway section in Zhenhai District is located in a marine depo-sitional plain; the entire line features soft soil. The geologic structure is shown in Fig. 12.5.

Table 12.3 Subgrade top surface rebound deflection test (unit: 0.01 mm)

Measuring point pile No.	Lane	Initial reading	Final reading	Rebound deflection	Initial reading	Final reading	Rebound deflection
K8+000	Left 1	87	80	14	275	260	30
K8+020	Left 1	185	177	16	148	142	12
K8+040	Left 1	343	334	18	166	153	26
K8+060	Left 1	369	358	22	442	431	22
K8+080	Left 1	304	300	8	445	405	80
K8+100	Left 1	689	677	24	158	144	28
K8+120	Left 1	828	817	22	160	147	26
K8+140	Left 1	137	126	22	278	258	40
K8+140	Left 2	107	98	18	111	99	24
K8+120	Left 2	348	330	36	51	38	26
K8+100	Left 2	372	364	16	137	124	26
K8+080	Left 2	39	30	18	734	723	22
K8+060	Left 2	119	105	28	151	139	24
K8+040	Left 2	161	151	20	127	115	24
K8+020	Left 2	167	153	28	192	180	24
K8+000	Left 2	149	139	20	117	109	16
Number of measuring points	Average rebound deflection value			Standard deviation		Representative rebound deflection value	
32	24.37			12.03		48.42	
Conclusion	The actual rebound deflection value is much smaller than the subgrade top surface rebound deflection control value						

The main geological condition is described as follows:

①₁ The artificial filling layer: slightly dense—loose; consists of macadam, block stone, clayey soil and some construction waste; constituents are extremely inhomogeneous.

①₂ The clay layer: grayish yellow, plastic; the lower part is soft plastic; thick-layer, medium—high compressibility; contains iron and manganese spot rendering, local little humus can be observed; the soil is inhomogeneous; smooth soil surface, luster response, no shaking response, high ductility, high dry strength. This layer is widely distributed in the entire line and is absent in the river section; the layer thickness is generally 0.5–3.60 m.

Before construction

After construction

Fig. 12.4 Comparison before and after subgrade formation

②₁ The mucky clay layer: gray, liquid-plastic; thick-layer; high compressibility; the soil is less homogeneous; local sludge; smooth soil surface; luster response, no

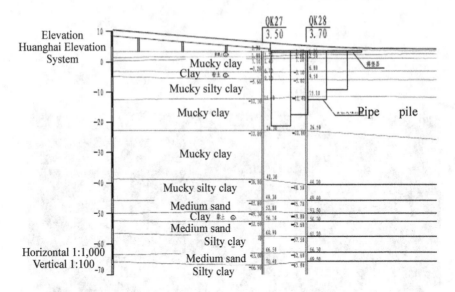

Fig. 12.5 Geologic structure

shaking response, high ductility; high dry strength; containing little shell debris. This layer is exposed in all drilled holes; the layer thickness is generally 17.7–25.40 m; the elevation of top plate is −2.20 to 1.45 m.

③₁ The silty sand, silt soil layer: gray, slightly dense—moderately dense, saturated, low dry strength, low ductility; particles are glued by clayey soil layer; local contains clayey soil layer. The layer thickness is generally 1.0–4.90 m; the elevation of top plate is −22.84 to −18.38 m.

③₃ The silty clay with sand layer: gray, liquid-plastic—slightly dense, saturated, low dry strength, low ductility; mainly clayey soil, silty soil, silt interbedding; local mucky soil. The layer thickness is generally 1.4–7.90 m; the elevation of top plate is −25.54 to −17.87 m.

④ The silty clay layer: gray, soft plastic, saturated, medium—high dry strength, medium ductility; with silt soil and silty sand thin layer; containing organic matter; high moisture content; local mucky soil. The thickness is generally 2.40–14.7 m; the elevation of the top plate is −34.18 to −21.18 m.

⑤ The silt layer: gray, slightly dense—moderately dense, saturated, low dry strength, low ductility, shaking response, local clayey soil layer. The thickness is generally 1.50–11.30 m; the elevation of top plate is −43.20 to −35.26 m.

⑥ The clay layer: the thickness is generally 0.5–7.75 m. This layer features deep soft soil; the moisture content is 40–56%; relatively large void ratio, but its mechanical properties are passable.

The main calculation indicators concerning the soil layers of the subgrade are shown in Tables 12.4 and 12.5.

Table 12.4 Physical and mechanical indicators concerning the subgrade soil layers

No.	Soil layer	Thickness	Saturated unit weight	Quick shear c	Quick shear internal friction angle	Consolidation quick shear internal friction angle
		m	kN/m^3	kPa	°	°
1.	Clay	0.5–3.60	18.5	27.3	13.2	15.0
2.	Muddy clay	17.7–25.40	18.0	6.8	4.8	8.5
3.	Silt	1.0–4.90	18.7	11.4	16.6	16.6
4.	Silty clay with sand	1.4–7.90	18.6	11.5	13.4	15.2
5.	Muddy-silty clay	2.40–14.7	18.5	14.6	10.0	13.5

Table 12.5 Soil layer E-P parameters (unit: kPa)

No.	Soil layer	E = 0 kPa	E = 50 kPa	E = 100 kPa	E = 200 kPa	E = 400 kPa
1.	Clay	0.884	0.841	0.813	0.773	0.719
2.	Muddy clay	1.336	1.212	1.142	1.038	0.917
3.	Silt	0.862	0.787	0.756	0.718	0.672
4.	Silty clay with sand	0.874	0.803	0.771	0.731	0.685
5.	Muddy-silty clay	1.037	0.957	0.912	0.858	0.789

The cross section arrangement of the subgrade is as follow: 4 m sidewalk + 3.5 m non-motor vehicle lane + 12 m motor vehicle lane + 5 m medial strip + 12 m motor vehicle lane + 3.5 m non-motor vehicle lane + 4 m sidewalk = 44 m.

The subgrade treatment mainly involves the bridgehead section. The subgrade filling is relatively high from 2 to 3.5 m. To prevent the bumps at bridgeheads and guarantee driving comfort, prestressed pipe piles and reinforced concrete frames are adopted for the soft subgrade treatment. Generally, the section filling is relatively low from 1.5 to 2 m. Preloading, presetting flip height and other measures are taken for the soft subgrade treatment.

The profile diagram and plane layout of the lower diaphragm plate treatment of the soft soil highway subgrade are shown in Figs. 12.6 and 12.7.

The bridgehead soft subgrade treatment in this project covers the reinforcement section and the transition section.

(1) The reinforcement section: reinforcement is made within the scope of 10.4 m from the bridgehead; the post-construction settlement in the pavement after treatment within the design service life is not larger than 10 cm to prevent the bumps at the bridgehead. The soft subgrade undergoes deep treatment with prestressed pipe piles; the pipe piles feature a square arrangement, and the pile

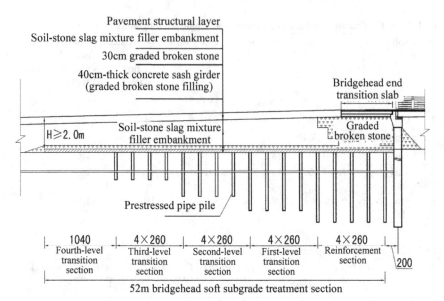

Fig. 12.6 Profile diagram of the lower diaphragm plate treatment of the soft soil highway subgrade (unit: cm)

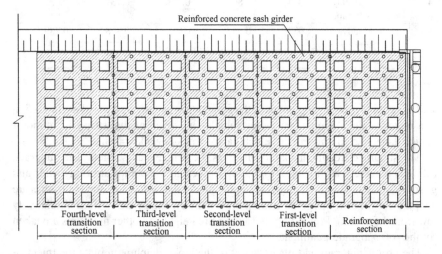

Fig. 12.7 Plane layout of the lower diaphragm plate treatment of the soft soil highway subgrade (unit: cm)

spacing is 2.6 m. The top of the pipe pile is provided with a 1.2 m-wide horizontal and vertical reinforced concrete sash girder, which guarantees the stability of the subgrade's forced deformation and fully utilizes the total force on the soil among the piles and pipe piles. The pile length at the reinforcement section is determined through theoretical calculation based on the situation in the soft subgrade at bridgehead.

(2) The transition section: at the general sections in this project, the soft subgrade treatment consists of reducing the height of the subgrade filing and adopting a preloading and presetting flip height to control the post-construction settlement at the general sections within 30 cm within the pavement design service life. The transition section is arranged between the reinforcement section for the bridge-head soft subgrade treatment and the general sections, and actions are taken to ensure that the transition of differential settlement meets the requirement that the gradual change ratio does not exceed 0.5%. For the section with adoption of prestressed pipe piles for soft subgrade treatment, its transition section is generally can be categorized into following three types: changeable pile length, changeable pile spacing and changeable pile length with changeable pile spacing. To guarantee the stability of the subgrade's forced deformation, the design of this project adopts reinforced concrete sash girders as the lower diaphragm plates for the subgrade. The longer the distance from the subgrade to bridgehead is, the lower its filling height is. To ensure that the subgrade produces the soil arch effect among the sash girders, increasing the pile spacing is inappropriate. Therefore, the pile spacing and sash girder size remain unchanged in the design, and transition is achieved by changing the pile length. The design covers four transition sections; the pile lengths at the first, second, and third transition sections are determined through theoretical calculations according to the situation in the soft subgrade at the bridgehead. No prestressed pipe piles are placed at the fourth transition section.

Index

© Springer Nature Singapore Pte Ltd. and Zhejiang University Press 2019
H. Zhu et al., *Controlling Differential Settlement of Highway Soft Soil Subgrade*,
SpringerBriefs in Applied Sciences and Technology,
https://doi.org/10.1007/978-981-13-0722-5

Printed in the United States
By Bookmasters